Lecture Notes in Artificial Intelligence 956

Subseries of Lecture Notes in Computer Science
Edited by J. G. Carbonell and J. Siekmann

Lecture Notes in Computer Science
Edited by G. Goos, J. Hartmanis and J. van Leeuwen

Springer

Berlin
Heidelberg
New York
Barcelona
Budapest
Hong Kong
London
Milan
Paris
Tokyo

Xin Yao (Ed.)

Progress in Evolutionary Computation

AI '93 and AI '94 Workshops
on Evolutionary Computation
Melbourne, Victoria, Australia, November 16, 1993
Armidale, NSW, Australia, November 21-22, 1994
Selected Papers

 Springer

Series Editors

Jaime G. Carbonell
School of Computer Science, Carnegie Mellon University
Pittsburgh, PA 15213-3891, USA

Jörg Siekmann
University of Saarland, German Research Center forAI (DFKI)
Stuhlsatzenhausweg 3, D-66123 Saarbrücken, Germany

Volume Editor

Xin Yao
Department of Computer Science, University College
The University of New South Wales, Australian Defence ForceAcademy
Canberra, ACT, Australia 2600

Cataloging-in-Publication Data applied for

Die Deutsche Bibliothek - CIP-Einheitsaufnahme

Progress in evolutionary computation : selected papers / AI '93
and AI '94 Workshops on Evolutionary Computation,
Melbourne, Victoria, Australia, November 16, 1993, Armidale,
NSW, Australia, November 21 - 22, 1994 / Xin Yao (ed.). -
Berlin ; Heidelberg ; New York : Springer, 1995
  (Lecture notes in computer science ; Vol. 956 : Lecture notes in
  artificial intelligence)
  ISBN 3-540-60154-6
NE: Yao, Xin [Hrsg.]; Workshops on Evolutionary Computation <1993 –
  1994, Melbourne; Armidale, New South Wales>; GT

CR Subject Classification (1991): I.2, G.1.6, G.3, J.4

ISBN 3-540-60154-6 Springer-Verlag Berlin Heidelberg New York

© Springer-Verlag Berlin Heidelberg 1995
Printed in Germany

Typesetting: Camera ready by author
SPIN 10486517   06/3142 – 5 4 3 2 1 0   Printed on acid-free paper

Preface

Evolutionary computation is the study of computational systems which use ideas and get inspirations from natural evolution and adaptation. The research in this field has progressed rapidly in recent years. This volume includes the papers presented at the AI'94 Workshop on Evolutionary Computation held in Armidale, New South Wales, Australia, 21–22 November 1994, and some of the papers presented at the AI'93 Workshop on Evolutionary Computation held in Melbourne, Victoria, Australia, 16 November 1993. Other papers presented at the workshops have been published in the journal *Informatica: An International Journal of Computing and Informatics, Volume 18, No. 4, December 1994*. They are:

D. T. Crosher, "The artificial evolution of adaptive processes," pp.377–386.

L. J. Fogel, D. B. Fogel and P. J. Angeline, "A preliminary investigation on extending evolutionary programming to include self-adaptation on finite state machines," pp.387–398.

T.Kido, K. Takagi and M. Nakanishi, "Analysis and comparisons of genetic algorithm, simulated annealing, tabu search and evolutionary combination algorithm," pp.399–410.

M. Tomita and T. Kido, "Sacrificial acts in single round prisoner's dilemma," pp.411–416.

J. Vaario, "From evolutionary computation to computational evolution," pp.417–434.

X. Yao and P. J. Darwen, "An experimental study of N-person iterated prisoner's dilemma games," pp.435–450.

The papers presented at the two workshops cover a wide variety of topics in the field of evolutionary computation, from evolutionary optimisation to evolutionary learning, from real-world applications to theoretical analysis. Although both workshops were held in Australia, more than half of the papers came from overseas (8 countries).

The success of a conference/workshop really depends on the close cooperation among participants, authors, reviewers, and organising committee members. I, as the committee chair, would like to take this opportunity to express my sincere thanks to all the authors, participants, and following members of the organising committee of the two workshops:

D. Abramson	Griffith University, Australia
E. Lewis	University College, University of New South Wales
	Australian Defence Force Academy, Australia
B. Marksjö	CSIRO Division of Building, Construction and Engineering
Z. Michalewicz	University of North Carolina — Charlotte, USA
H. B. Penfold	University of Newcastle, Australia
O. de Vel	James Cook University of North Queensland, Australia
K. P. Wong	University of Western Australia, Australia

and the following reviewers:

D. Abramson	A. N. Burkitt	P. J. Darwen	I. Davidson
D. Fraser	C. Jones	E. Lewis	B. Litow
B. Marksjö	Z. Michalewicz	H. B. Penfold	D. Sier
G. S. Trinidad	C. P. Tsang	A. Varsek	O. de Vel
P. Whigham	K. P. Wong	X. Wu	A. Zomaya

Finally, I would like to thank Professor J. Siekmann for his support and cooperation in editing this volume.

May 1995   Xin Yao

Table of Contents

The Effect of Function Noise on GP Efficiency

Jack Y.B. Lee and P.C. Wong

Advanced Network Systems Laboratory
Department of Information Engineering
The Chinese University of Hong Kong
Hongkong
yblee@ie.cuhk.hk

Abstract. Genetic Programming (GP) has been applied to many problems and there are indications [1,2,3] that GP is potentially useful in evolving algorithms for problem solving. This paper investigates one problem with algorithmic evolution using GP - *Function Noise*. We show that the performance of GP could be severely degraded even in the presence of minor noise in the GP functions. We investigated two counter-noise schemes, *Multi-Sampling Function* and *Multi-Testcases*. We show that the *Multi-Sampling Function* scheme can reduce the effect of noise in a predictable way while the *Multi-Testcases* scheme evolves radically different program structures to avoid the effect of noise. Essentially, the two schemes lead the GP to evolve into different "approaches" to solving the same problem.

1 Introduction

1.1 Genetic Programming

The Genetic Programming (GP) paradigm developed by Koza [1] which evolves hierarchical tree-based LISP like programs to solve a surprisingly wide variety of problems hold much promise among the Evolutionary Algorithms paradigms. In particular, the structural advantage of GP holds strong promise in evolving algorithms for problem solving. Many examples have been demonstrated by Koza [1], Kinnear [2,3], and Tackett [5], showing that evolving algorithms including decisions, iterations and recursions are possible.

Koza characterizes a GP problem into five steps [1]:

1. Define the Terminal set,
2. Define the Function set,
3. Define the Fitness measure,
4. Select Control parameters,
5. Define Termination and result designation.

Then an initial population is randomly generated and the evolution process is started. At each generation, the quality of every individual in the population is evaluated using the fitness measure defined in Step 3 above. Fitter individuals are given more chance to reproduce offsprings into the next generation. The population hopefully will evolve into better solutions. The evolution process is terminated when either the maximum number of runs are reached or the termination criterion defined in Step 5 is satisfied.

1.2 Noise

Noise is common in real-world applications, especially when interfacing to the physical-world is involved. For example, real-world sensors and detectors are all susceptible to various kinds of noise. If the algorithm is evolved by GP in a simulated environment without accounting for noise, the performance of the evolved algorithm in such real-world applications is highly questionable.

Despite the importance of noise in the application of GAs and GPs, there has been very little study in this area. Most study and analysis of noise in GAs deal with noise which are implicit in GA, notably due to the finite population size limitation. This paper addresses noise resulting from the problem domain, instead of from the GA domain.

Consider the five steps of GP described in Section 1.1 above; the first three steps, namely defining terminal set, defining function set and defining fitness measure, are the main abstractions of the real-world problem. If the abstracted object is noisy, e.g. a sensor, the corresponding GP representation will have direct counterparts reflecting the noise in the GP algorithm. For example, if a function in the function set abstracts a real-world sensor and the sensor is subjected to noise, the GP function should model the same noise.

1.3 Artificial Ant

A well-known sample problem is the Artificial Ant problem [1]. In this problem, the ant's movement is controlled by the GP evolved program, which includes the decision function *'IfFoodAhead'*. The goal of the Artificial Ant is to find and *'eat'* as many food pellets as possible in a limited time. Table 1 shows the function set and terminal set for the Artificial Ant problem, and Figure 1 shows the map of the Santa Fe Trail.

Table 1. Artificial Ant

Function Set	IfFoodAhead(2), ProgN2(2), ProgN3(3)
Terminal Set	Left, Right, Move

The Artificial Ant problem falls into a special class of problems which we termed *Iterative Control Problems* (ICPs). These problems have the following characteristics:

1. A single algorithm is used to control the system's behaviour, in this case the ant's behaviour.

Fig. 1. The Santa Fe Trail

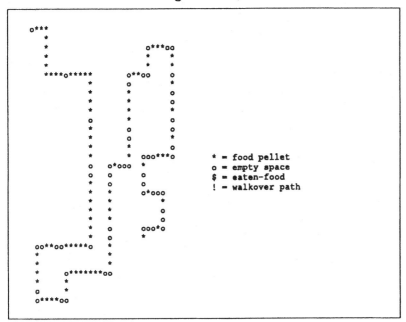

* = food pellet
o = empty space
$ = eaten-food
! = walkover path

2. A run of the system requires multiple executions of the algorithm.
3. The next state of the system depends on the current state of the system.

Other examples include *Robotics control* [6,7,8] and *Obstacle avoidance* [9,10,11] problems. To empirically investigate the effect of noise on these kind of problems, we have chosen the Artificial Ant Problem as a testbed problem.

2   Problem Formulation

In the original GP Artificial Ant problem [1], 3 functions and 3 terminals are used as shown in Table 1 above. The function *IfFoodAhead* takes two parameters and evaluates the first parameter if there is food one step ahead and evaluates the second parameter otherwise. In the original setting, the *IfFoodAhead* function always makes the correct decision. In other words, no noise is present in executing the function. The remaining two functions *ProgN2* and *ProgN3* are used only as connective functions. See Koza [1] for details.

2.1   Noise

To introduce noise into the GP function, a modified *IfFoodAhead* function which occasionally makes incorrect decisions is created. The level of noise is controlled by a single parameter, namely *NOISELEVEL*. The procedure to execute the function is as follows:

- Step 1. Determine the true result using the original method.
- Step 2. Produce a random number X between 0 and 1.
- Step 3. If X is equal or greater than $NOISELEVEL$, the 'correct' result is returned. 'Incorrect' result is returned otherwise.

Note that the random number in Step 2 is generated using the Unix C Library function *random()* [4]. $NOISELEVEL$ is selectable from 0 to 1.

The $NOISELEVEL$ parameter is symmetric about 0.5. That is, a noise level of 1 (always return opposite results) or 0 (always return true results) implies no noise while a noise level of 0.5 implies the function is blinded by noise, that is, the *IfFoodAhead* function provides no information at all.

2.2 Experiments

In the first experiments, the parameters are set after Koza [1] as shown in Table 2.

Table 2. Simulation Parameters

Parameter	Value
Objective	Find a computer program to control an artificial ant so that it can find all 89 pieces of food located on the Santa Fe trail.
Terminal Set	(LEFT), (RIGHT), (MOVE)
Function Set	Noisy IfFoodAhead, ProgN2, ProgN3
Fitness Cases	One fitness case
Raw Fitness	Number of pieces of food picked up before the ant times out with 400 operations.
Standard Fitness	Total number of pieces of food (89) minus raw fitness plus 0.1 times the total number of nodes and leaves in the evolved program (Parsimony factor of 0.1).
Hits	Same as Raw Fitness
Population	500
Generations	500

With these parameters, runs with noise levels 0.00, 0.02, 0.10, 0.30, 0.50, and 0.95 are performed. The results are obtained by averaging multiple runs of the same parameters. The comparative results of Best Hits, Average Hits of the various noise levels are shown in Figure 2 and 3 respectively.

2.3 Results

Best Hits Best Hits is defined as the value of the fitness of the best individual in the particular generation. The Best Hits figure shows how good is the best individual in any generation.

From Figure 2, we observe that the best hits is reduced by approximately 20% (noise level 0.02) to 40% (noise level 0.50) in the presence of noise. We note that noise level of 0.50 implies a blinded Artificial Ant, the runs with lower noise levels (e.g. 0.30, 0.10 and 0.02) perform only marginally better than a blinded Artificial Ant.

Fig. 2. Best Hits for Plain-Runs

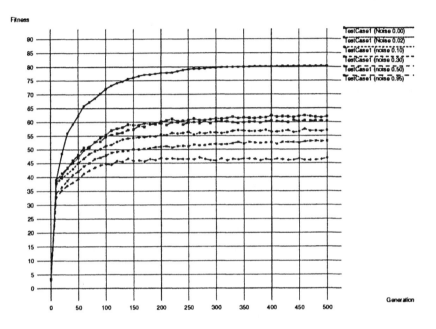

Average Hits Average Hits is defined as the average fitness of all individuals in the particular generation. The Average Hits figure shows how good the whole population is taken together.

From Figure 3, it is observed that the (population) average hits is reduced by over 45% even in the presence of noise level as low as 0.02. And for all the noise levels, ranging from 0.02 to 0.5 to 0.95, the average hits varies within the close range of 18 to 27 after 200 generations. As expected, the worse runs are that with noise level of 0.5, which is the same as blinding the ant. Therefore, with reference to the 0.5 noise level runs, the other runs with various noise levels are only marginally better in terms of average hits.

2.4 Discussions

It is surprising that even with a noise level of 0.02, the performance of GP is greatly degraded. And within different noise levels, the difference in performance

Fig. 3. Average Hits for Plain-Runs

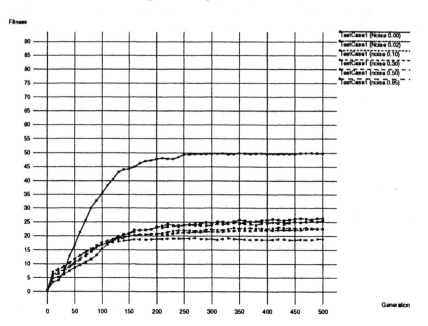

is lower than expected. The effect of having a noise level greater than 0.5 is the same as reversing the meaning of the noisy function. For this problem, the function *IfFoodAhead* with noise level x will behave as IfFoodNotAhead of noise level $(1.0 - x)$.

From both figures, the performance of the GP evolved Artificial Ant program is clearly significantly degraded. This can be accounted for by two reasons:

1. The evolved program is being executed repeatedly until time-out or all food are eaten, so the probability of having a noise-induced error is multiplied by the many evaluations of the noisy function during a single fitness evaluation. Furthermore, the food pellets are located in a trail-like pattern (cf. property 3 of ICPs). Once the ant misses a food pellet due to noise, it could easily wander off the food trail. (see Figure 4).

2. In the presence of noise, the evaluated fitness is highly susceptible to noise. The fitness value obtained may not reflect the *true* fitness of the individual. A single algorithm could produce very different fitnesses in different runs. In this way, the evolution may be misled by non-representative fitness values and fails to find the optimum algorithm.

In the following sections, we investigate two schemes in attempt to counter the impact of noise on the GP performance.

Fig. 4. A run with noise level 0.02

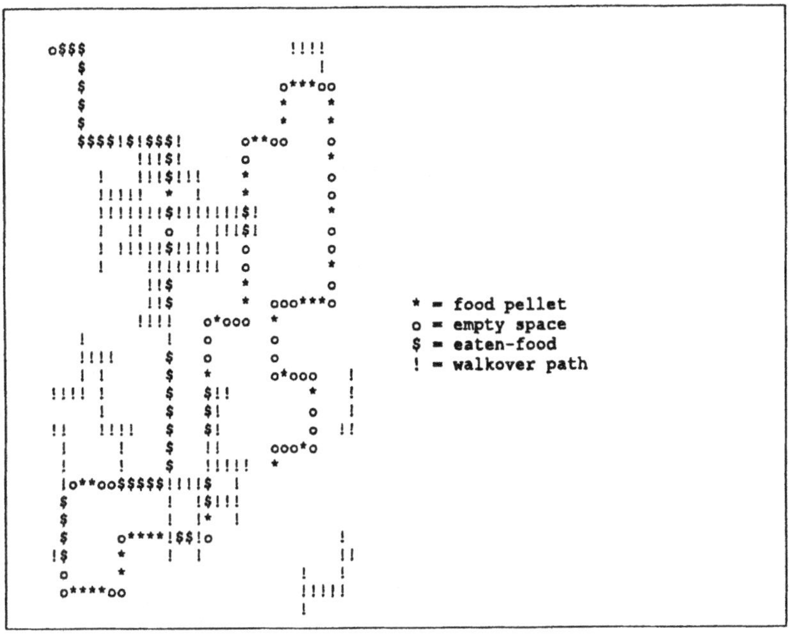

```
o$$$                    !!!!
 $                       !
 $                      o***oo
 $                       *     *
 $                       *     *
 $$$$!$!$$$!      o**oo   o
    !!!$!         o       *
   !  !!!$!!!     *       o
  !!!!!  *  !     *       o
  !!!!!!!$!!!!!!!$!        *
   !  !!  o  ! !!!$!       o
   ! !!!!!$!!!!!  o        o
   !    !!!!!!!!  o        *
      !!$         *        o
      !!$         *   ooo***o      * = food pellet
     !!!!    o*ooo   *             o = empty space
   !     !     o      o            $ = eaten-food
   !!!!        $      o            ! = walkover path
   ! !         $      *   o*ooo   !
 !!!! !        $   $!!      *     !
   1           $   $!       o     !
 !! !!!!       $   $!       o    !!
  !   !        $   !!    ooo*o
  !     !      $   !!!!!   *
  !o**oo$$$$!!!!$   !
  $           !  !$!!!
  $           !  !*   !
  $      o****!$$!o               !
 !$      *    !  !               !!
  o      *                     !   !
  o***oo                       !!!!!
                                !
```

3 Counter-Noise Schemes

Two schemes are investigated here, namely *Multi-Sampling Function* and *Multi-Testcases*.

3.1 Multi-Sampling Function

In this problem, the single noise source comes from the noisy *IfFoodAhead* function. Assuming the function can be modified in such a way that instead of sampling (evaluating) the environment once, it samples the environment n times, where $n \geq 1$. The final result of the function is then determined by simple majority voting.

Specifically, assume that $NOISELEVEL = L$, and the number of samples equals n. The noise level of the original single-sampling function is:

$$N_0 = L \tag{1}$$

We define the *effective* noise level N_e as the equivalent noise level of the resulting multi-sampling function, which is given by:

$$N_e = \sum_{x=(n+1)/2}^{n} {}_nC_x N_0^x (1 - N_0)^{(n-x)} \qquad \text{assume } n \text{ is odd} \tag{2}$$

Table 3 shows the effective noise equivalent for $n = 11$ and Figure 5 shows the effective noise level N_e versus the original noise level N_0 using $n = 3, 5, 11$, and 17 respectively.

Table 3.

Original Noise Level N_0	Effective Noise Level N_e for $n = 11$
0.00	0.00
0.02	2.71207E-08
0.10	0.000295706
0.30	0.078224791
0.50	0.50
0.95	0.999994199

Fig. 5. Noise Level N_0 vs Effective Noise Level N_e for $n = 3, 5, 11$, and 17.

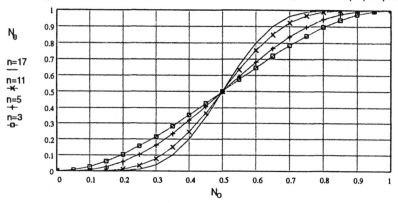

According to Figure 5 above, the effective noise level N_e decreases as n increases for a given noise level N_0. Secondly, the amount of noise reduction decreases as the noise level approaches the noise level $N_0 = 0.5$ from either side of the curve.

Experiments and Results To test the idea experimentally, the same sequence of experiments as described in Section 2 is performed again with the modified *multi-sampling IfFoodAhead* function (*IfMultiFoodAhead*), using $n = 11$. The results are shown in Figure 6 and 7.

The experimental results agree very well with the theoretical calculations given above and the results in Section 2. The performance of the runs are greatly improved and the improvement is closely predicted by Equation 2 given in Section 3.1. For example, for $N_0 = 0.3$, the corresponding N_e is 0.078 (Table 3),

Fig. 6. Best Hits for Multi-Sampling Scheme

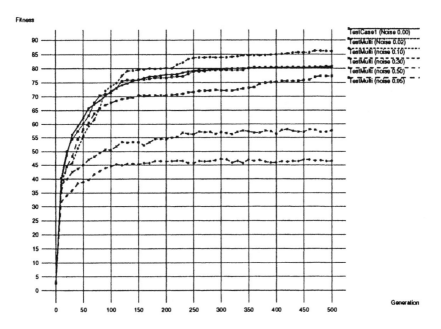

Fig. 7. Average Hits for Multi-Sampling Scheme

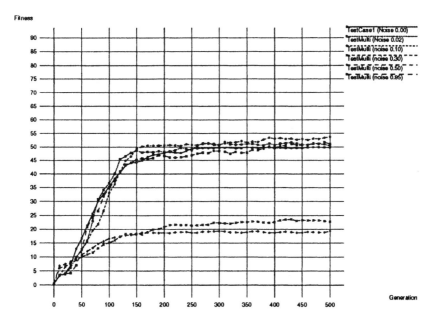

with Best Hits around 57 according to Figure 6. This is very close to the Best Hits of $N_e = 0.1$ in Figure 2. Also, for the special case of $N_0 = 0.5$, where N_e is equivalent to N_0 for all n, both plain runs and multi-sampling runs are equivalent (refer to Figure 2 and 6).

Discussions The above results show that the Multi-Sampling Function scheme can indeed improve the performance of the genetic search. The improvement is most significant at low noise levels (see Figure 5). However, the assumption that the noisy function can be sampled multiple times for a single evaluation may have problems or even impossible in certain applications. On the other hand, the cost (time and otherwise) for multiple-sampling is problem dependent.

3.2   Multi-Testcases

As discussed in Section 2.4 (2), in the presence of function noise, single fitness evaluation may not provide adequately accurate information to guide the genetic search. One evaluation for each individual may not reflect the true fitness for that particular individual. By using the Multi-Testcases scheme, instead of performing a single run to determine the fitness, the fitness of an individual in each generation is averaged over N independent evaluations, where $N \geq 1$.

Define F_n as the fitness of the nth evaluation, the Multi-Testcases fitness F_m is given by:

$$F_m = \frac{1}{N} \sum_{n=1}^{N} F_n \qquad \text{where } N \geq 1 \qquad (3)$$

In this way, the fitness of each individual should be more accurate and therefore provide better guidance for the genetic search.

Experimental Results To test the idea experimentally, the same sequence of experiments as described in Section 2 is run again using $N = 10$. The results are shown in Figure 8 and 9.

Both the Best Hits (Fig. 8) and Average Hits (Fig. 9) measures show that the performance under various noise levels are very close and shows no significant difference across the whole noise level spectrum from 0.02 to 0.95. The performance of the genetic search appear to be *independent* of noise level.

Discussions In all cases, the Average Hits is superior to runs without multiple testcases, while the Best Hits performance compare favourably with plain runs at high noise levels. This is because the plain runs' Best Hits performance deteriorates quickly with increasing noise (Fig.2) while the runs with multiple testcases stay relatively constant across various noise levels (Fig.8).

This surprising result seems to indicate that the Multi-Testcases scheme has not achieved the original hope. However, the Average Complexity measures, defined as the total number of nodes and terminals, reveal another story (Fig. 10, 11).

Fig. 8. Best Hits for Multi-Testcases Scheme

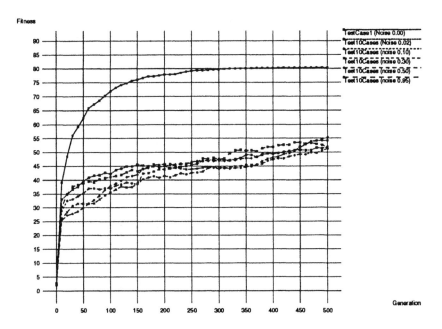

Fig. 9. Average Hits for Multi-Testcases Scheme

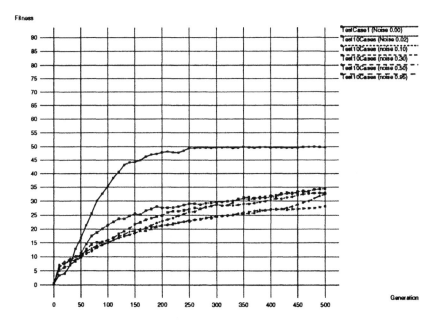

Fig. 10. Average Complexity for Noise Level of 0.02

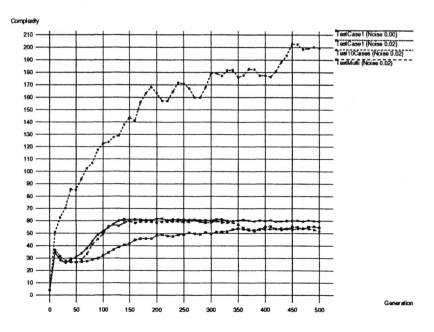

Fig. 11. Average Complexity for Noise Level of 0.30

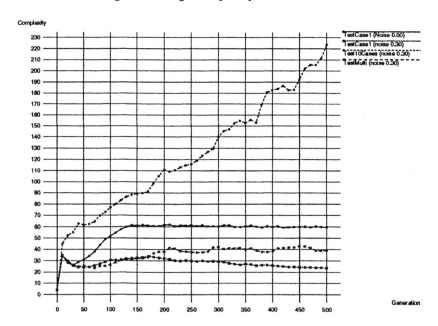

3.3 Multi-Sampling Function vs Multi-Testcases

The Average Complexity measures (Fig. 10, 11) reveal that the average complexity of the individuals evolved using the Multi-Testcases scheme is significantly higher than all the runs in Section 2 (original setting) and Section 3.1 (Multi-Sampling Function Scheme). The difference is so great that the evolved population is very different in structure (species) than those in plain-runs and runs under the Multi-Sampling Function scheme.

Inspecting the population individuals confirms that the population under the Multi-Testcases scheme evolved using entirely different *"approaches"* to solve the Noisy Artificial Ant problem. Specifically, the runs in Section 2 and 3.1 are evolving towards programs than can *follow* the Santa Fe Trail as closely as possible. On the contrary, the population here evolved towards programs that do not follow the Santa Fe Trail, but rather use a *'sweeping'* techniques to catch as many food pellets as possible. Figure 12 and 13 show two such evolved program. Note that the regularity of the Santa Fe Trail is exploited to achieve fairy good performance regardless of the noise level. Another observation is that the best individual evolved (as in Figure 12 and 13) do not use the noisy *IfFoodAhead* function at all. They move in a *deterministic* manner.

Fig. 12. A Multi-Testcases run with noise level 0.30

```
o$$$!   !   !   !   !   !   !
  *!!!!!   !   !   !   !   !
  *!   !!!!!   !   !   !***!o
  *!   !   !!!!!   !   $   !*
  *!   !   !   !!!!!   $   !*
  *$**o$***$   !   !$$!!   !o
  !   !   $   !   !   !!!!!*
  !   !   $   !   $   !   !o      * = food pellet
!!!!!   $   !   $   !   !o        o = empty space
!   !!!!!   $   !   $   !   !*     $ = eaten-food
!   !   !!!!!   !   $   !   !o     ! = walkover path
!   !   !   $!!!!   !   !o
!   !   !   $   !!!!!   !   !*
!   !   !   $   !   $!!!!   !o
!   !   !   $   !   $   o!!$$$o
!   !   !   !   !*oo!   *!   !!!!
!!!!!   !   !   !   !   o!   !
!   !!!!!   $   !   !   o!   !
!   !   !!!!$   $   !   o$oool    Ant Info: hits=51
!   !   !   $!!!$   !   !   *!    Program : (progn2 (progn2 (move)
!   !   !   $   $!!!!   !   o!      (move)) (progn3 (progn2
!   !   !   $   $   !!!!!   o!      (progn3 (move)   (move)
!   !   !   $   !   o!!$!!          (left)) (progn2 (move)
!   !   !   $   !   !   *!   !!!!    (move))) (progn2 (move)
!!!$$oo*$***!   $   !   !   !       (move)) (progn2 (progn2
!*  !!!!!   !   $   !   !   !       (move)   (move)) (progn2
!*  !   !!!!!   $   !   !   !       (move) (right)))))
!*  !   o$***$$$!!   !   !   !
!*  !   *!   !   !!!!!   !   !
!o  !   *!   !   !   !!!!!   !
!o**$*oo!   !   !   !   !!!!!
!   !   !   !   !   !   !   !!!!
```

According to the results from Section 2.3, the fitness of path-following individuals are severely degraded in the presence of noise. Therefore, under the

14

Fig. 13. A Multi-Testcases run with noise level 0.50

```
o$$$!       !       !
  *!!!!!!!!!       !
  *!        !!!!!!!!!!   o***oo
  *!        !       !!!!$    *
 !!!$!      !       !    *    !$!!
  *$$$!$$$$$        !**oo     !o
    !     $!!!!!!!!!        !*
    !     $         $!!!!!!!!!o          * = food pellet
 !!!!!    $         $         !!!!       o = empty space
   !!!!!!!!$        $         !*         $ = eaten-food
    !     !!!!!!!!!$          !o         ! = walkover path
    !     $         !!!!!!!!!!o
 !!!!!    $         !         !$!!
   !!!!!!!!$        $         !o
    !     $!!!!!!!$  ooo**$o
    !       !  o*oo!!!$!!!!!
 !!!!!      !   o    !   o    !!!!
   !!!!!!!!$    o    !   o    !
    !     $!!!$!!!!  o*oool
    !     $    *     !!!!!!!!$!     Ant Info: hits=48
 !!!!!    $    *     !    o!!!!     Program : (progn3 (move) (progn3
   !!!!!!!!$   *     !     o!                (move)   (move)  (move))
    !     $!!!!!!!!  ooo*o!                  (progn3 (left) (progn3
    !     $    o    !!!$!!!!!                 (move) (move) (move))
 !!!$$oo*****!   *    !        !!!!          (progn3 (right) (progn3
  *  !!!!!!!!!   *    !        !              (move) (move) (move))
  *  !     !!!!$!!!!          !               (move))))
  *  ! o****$**oo      !!!!!!!!!
 !$!!!  *    !         !        !!!!
  o  !!!$!!!!!         !        !
  o**$*oo    !!!!!!!!!         !
     !       !       !!!!!!!!!!
```

Multi-Testcases scheme, the averaged fitness of the path-following individuals tend to be worse than the others.

The observation here is that in the Multi-Testcases Scheme, the genetic search tends to evolve more *robust* programs like those in Figure 12 and 13 above, which do not use the error-prone noisy *IfFoodAhead* function. On the other hand, the Best Hits performance is reduced as a trade-off for robustness.

4   Conclusions

While Genetic Programming is promising in solving complex problems by evolving programs and algorithms, the high sensitivity to function noise poses challenges when GP evolved programs are put into real-world applications, where noise is inevitable.

This paper investigated the effect of function noise on the performance of GP and showed that GP evolved algorithms are highly sensitive to function noise. Due to the iterative nature of some evolved programs and algorithms, the performance can be significantly degraded even in the presence of minor function noise.

To counter the function noise problem, two schemes, *Multi-Sampling Function* and *Multi-Testcases* are investigated and we found that they evolve radically different programs and algorithms for solving the same problem. In this particular case, the Multi-Sampling Function scheme gives better performances provided that the stated assumptions are satisfied. The Multi-Testcases scheme performs well only at very high noise levels while being more robust to noise.

The results here show that the direction of a genetic search can be highly *biased* by the fitness evaluation method employed. Although the goal of the Multi-Sampling Function and Multi-Testcases Schemes are similar, the resulting evolution could be biased into radically different directions.

The results given here are only preliminary but have already shown the importance of noise on the efficiency of a certain class of problems. Besides the *Function Noise* investigated here, noise in the terminals (*Terminal Noise*) and fitness function (*Fitness Noise*) also need investigations. These issues will become more important when GP evolved algorithms are applied to real-world applications, where noise is the norm, not the exception.

5 Acknowledgements

All the experiments in this paper is run using the *Geppetto Genetic Programming System* written by Dave Glowacki. Elitism is added by the authors. We wish to thanks the anonymous reviewers for their constructive comments. This paper is supported in part by the *Postgraduate Student Grants for Overseas Academic Activities* of *The Chinese University of Hong Kong*.

References

1. J. Koza. Genetic Programming. MIT Press. Cambridge, MA.
2. Kinnear, Kenneth E. Jr. Evolving a sort: Lessons in genetic programming. 1993 IEEE International Conference on Neural Networks, San Francisco. Piscataway, NJ: IEEE 1993. Volume 2. Pages 881-888. 1993a.
3. Kinnear, Kenneth E. Jr. Generality and difficulty in genetic programming: Evolving a sort. In Forrest, Stephanie (editor). Proceedings of the Fifth International Conference on Genetic Algorithms. San Mateo, CA: Morgan Kaufmann Publishers Inc. Pages 287-294. 1993b.
4. Unix Programmer's Guide. Online man pages.
5. Tackett, Walter Alden. Genetic programming for feature discovery and image discrimination. In Forrest, Stephanie (editor). Proceedings of the Fifth International Conference on Genetic Algorithms. San Mateo, CA: Morgan Kaufmann Publishers Inc. Pages 303-309. 1993a.
6. The genetic planner: The automatic generation of plans for a mobile robot via genetic programming. Proceedings of the Eighth IEEE International Symposium on Intelligent Control. Pages 190-195.
7. Unix Programmer's Guide. Online man pages.
8. Handley, Simon. The automatic generation of plans for a mobile robot via genetic programming with automatically defined functions. In Proceedings of the 1993 International Simulation Technology Multiconference (SimTec '93).
9. Handley, Simon. The automatic generation of plans for a mobile robot via genetic programming with automatically defined functions. In Kinnear, Kenneth E. Jr. (editor). Advances in Genetic Programming. Cambridge: The MIT Press 1994.

10. Reynolds, Craig W. An evolved vision-based behavioral model of coordinated group motion. In Meyer, Jean-Arcady, Roitblat, Herbert L. and Wilson, Stewart W. (editors). From Animals to Animats 2: Proceedings of the Second International Conference on Simulation of Adaptive Behavior. Cambridge, MA: The MIT Press. Pages 384-392.

11. Reynolds, Craig W. Evolution of obstacle avoidance behavior: Using noise to promote robust solutions. In Kinnear, Kenneth E. Jr. (editor). Advances in Genetic Programming. Cambridge: The MIT Press.

12. Reynolds, Craig W. An evolved vision-based model of obstacle avoidance behavior. In Langton, Christopher (editor). Artificial Life III.

Genetic Approaches to Learning Recursive Relations

P.A.Whigham and R.I.McKay

Department of Computer Science, University College, University of New South Wales
Australian Defence Force Academy Canberra ACT 2600 AUSTRALIA Email:
paw@csadfa.cs.adfa.oz.au Phone: 06-268-8182 Fax: 06-268-8581

Abstract. The genetic programming (GP) paradigm is a new approach to inductively forming programs that describe a particular problem. The use of *natural selection* based on a *fitness function* for reproduction of the program population has allowed many problems to be solved that require a non-fixed representation. Issues of typing and language forms within the genetic programming paradigm are discussed. The recursive nature of many geospatial [1] problems leads to a study of learning recursive definitions in a subset of a functional language. The inadequacy of GP to create recursive definitions is argued, and a class of problems hypothesised that are difficult for genetic approaches. Operations from the field of *Inductive Logic Programming*, such as the **V** and **W** operators, are shown to have analogies with GP crossover but are able to handle some recursive definitions. Applying a genetic approach to ILP operators is proposed as one approach to learning recursive relations.

Topic: Machine Learning
Keyword List: Machine Learning, Inductive Logic Programming
                 Genetic Programming

1 Introduction

The Genetic Programming paradigm (GP) has received some attention lately as a form of adaptive learning [4]. The technique is based upon the genetic algorithm (GA), [3], which exploits the process of natural selection based on a fitness measure to breed a population that improves over time. The ability of GA's to efficiently search large conceptual spaces makes them suitable for the discovery and induction of generalisations from a data set. A summary of the genetic programming paradigm may be found in [11].

This paper describes the application of the genetic programming paradigm to a simple recursive problem in a subset of the functional language LISP. Recursive descriptions are of interest as an examination of the spatial properties

---

[1] Problems where the accepted model involves spatial descriptions normally used in geographical discourse

of natural processes indicates that many processes are described *most clearly* by using recursive statements. The negative results that occur suggest that GP is not suitable for discovering recursive definitions, and therefore other techniques are required to breed computer programs that have a recursive structure.

Our primary motivation is to develop techniques for learning geospatial models arising in resource management domains. These spatial relationships are of relatively low complexity - certainly comparable with previous successful applications of the GP paradigm - but are atypical in often requiring a *recursive representation*.

The field of Inductive Logic Programming (ILP) is introduced, and a number of operators from ILP are shown to perform the appropriate (yet non-deterministic) operations which allow the detection of recursive descriptions. The non-determinism is taken advantage of by using GP's to breed a number of alternatives that may be further explored in a parallel (broad beam) search fashion. A field of research is described that brings together the operations of Genetic Programming and Inductive Logic Programming to create generalisations in a formal language.

2   The Genetic Program

The Genetic Programming Paradigm was first introduced by Koza [7] as a way to extend the genetic algorithm to hierarchical structures. The main criterion for using GP involves defining the *terminals and non-terminals* that are used to compose the program. The non-terminals are referred to as *functions*. These functions usually include logical operators, mathematical functions and problem-specific functions. The terminals represent actual values which may be used as *arguments* to the functions.

The GP paradigm states that the terminal set and function set should be selected to satisfy the requirements of *closure* and *sufficiency*.

2.1   Closure and Sufficiency

Closure is used to indicate that the function set should be well defined for any combination of arguments. This allows any two points in a program to be *crossed over* by swapping their program structures at these points in the program tree. Koza [4], discussing the issue of *closure*, states:

> "Closure can be achieved in a straightforward way for the vast majority of problems merely by careful handling of a small number of situations."

Problems that require typing are handled by constraining the syntactic structure of the resultant programs [5]. In particular, when a crossover point is selected, the second point for crossover must match the syntactic type of the first point. This ensures that only *legal syntactic programs* are created when 2 programs swap components [4]. Sufficiency refers to the fact that the GP constructs must have the expressive power to actually solve the problem being posed.

2.2 Learning Recursive Functions

Although Koza has used restrictions to allow typed functions with the GP paradigm, the issues of creating recursive definitions has been only briefly considered [6]. Our initial interest in this area stemmed from learning about geographic relations. The importance of inductively creating descriptions of spatial properties is obvious from the amount of spatial information that exists and continues to be generated. A procedure that creates generalisations of this spatial information may be used for classification, explanation and prediction. Spatial properties in natural resource domains are often best described by a recursive language. For example, the *wetness* of a particular location may be defined in terms of the *wetness* of adjacent locations. Similarly, the population density of an animal will generally be influenced by the population (of the same animal) at nearby locations. In section 3 an example involving *nitrate concentrations* will be described. An inductive approach that does not allow recursion will miss the concise and fundamental descriptions that capture these underlying processes.

At a theoretical level, the ability to learn recursive definitions opens the possibility for learning self-referential concepts and identifying isomorphic structures.

A Simple Subset of LISP We attempted to learn the simple recursive definition of *member(X, Y)* using a typed subset of LISP. LISP [1] is a functional language that, in its pure form, uses no variable assignment and all arguments are passed *as value parameters*. Given a number of positive and negative examples of *member*, the program used the non-terminals *CAR, CDR, EQ, ATOM and MEMBER* to build a recursive definition of member. The terminals *X and Y* were used as the arguments to these functions. The following typing constraints were applied, where ATOM is a LISP atomic value, and LIST represents the simple data structure of a list of atoms:

GP Terminals (arguments)	
Argument	Type
T (true)	ATOM
NIL (false)	ATOM
X (var)	ATOM
Y (var)	LIST or ATOM

GP Non-terminals (functions)		
Function	Arg. Types	Return Type
EQ	(ATOM,ATOM)	ATOM (T or NIL)
CAR	(LIST)	ATOM
CDR	(Y)	LIST or ATOM (NIL)
COND	(ATOM,ATOM,ATOM)	ATOM
ATOM	(Y)	ATOM (T or NIL)
MEMBER	(X,LIST)	ATOM (T or NIL)

Note that the functions EQ,CAR,CDR,COND and ATOM represent the standard LISP functions of the same name.

The fitness measure for each program was based on the number of true and false member examples that were correctly classified. A *raw fitness* of zero implied that all examples were correctly determined. As programs could be potentially created that recursed indefinitely, a stack limit of 40 calls was imposed. As all example membership tests had list lengths of less than 40 this limit seemed adequate. A program that caused a stack overflow returned **false** for the member function.

Results: Learning Recursive Member The input file for learning the member definition consisted of examples which showed various combinations of placement for the element X and its position in the list Y, such as

```
MEMBER (1,( 3 )) :- NIL
MEMBER (1,( 2 3 8 9 )) :- NIL
MEMBER (1,( 1 )) :- T
MEMBER (1,( 3 5 7 1 )) :- T
```

The examples were split evenly amongst 10 failures and 10
true values. Of these,
    1 example had the first element as the member,
    7 had the second element as the member
    2 had the member spread towards the end of the list.

The tests were performed with the following GP settings:

Genetic Programming Controls	
GENERATIONS	51
POPULATION SIZE	2000
INITIAL POP LOW	2
INITIAL POP HIGH	5
DEPTH FULL	250
DEPTH RANDOM	250
CROSSOVER	1800
REPRODUCTION	200
CROSSOVER-INT	90%

We found that the GP could not sucessfully find the definition of member that was required, although it satisfied the conditions of closure and sufficiency. Typically the trivial member predicate which defined member as the first list element was found, however the general predicate using the recursive call of *MEMBER(X,CDR(Y))* was not found. This was not to say that the *MEMBER(X,CDR(Y))* call was not included in the population, rather that this call requires the previous test to ensure that Y is not an ATOM. As such, when the recursive *MEMBER* function was used in the population it inevitably led to an infinite recursion, with the subsequent failure of the program. For example, the test for second element membership:

```
MEMBER(X,Y) :-  EQ( X  CAR( CDR( Y ) ) )
```

was found, which had a raw fitness of 5. Hence it failed on the 2 tests with the member further than the second element, the example with the member as the first element, and the 2 NIL tests where only one element existed in the Y list. The best program that was created had a raw fitness of 4:

```
MEMBER(X,Y):-
          COND( EQ( X  CAR( Y )) -> T : EQ( CAR( CDR( Y )))
```

which effectively represents a test of membership for the first and second positions of the list Y. No program found the correct solution, although the components of the population included elements that could be used to construct the correct program solution.

Discussion of Negative Results Although the possible crossover combinations were constrained by limiting the types of each functional argument, the program failed to find the proper recursive definition. This appears to be because adding partial solutions (say those that find the first-element member) to further solutions does not lead towards finding the recursive (and therefore general) solution. The *fitness function* in no way supports the discovery of these generalisations, because the GP paradigm does not consider whether any partial solution is either *too general or specific*. There is also no mechanism to consider the internal structure of the partial solutions that have been created, which must surely be necessary when a complex description is attempted.

In general we can account for this failure as a separation of the syntax and semantics of the analysis. The GP paradigm uses a syntactic approach to building programs, without knowledge of the interaction and usefulness of each component of a potential solution. Hence, a program that fails because it uses a recursive call to *MEMBER* will be unlikely to promote this construct, although it required only the addition of a stopping condition to halt the recursion when the list *did not contain the member X*. This type of problem does not normally occur with a fixed-length genetic algorithm as recursion is not normally possible with the representation, or its stopping conditions are controlled in a higher-level function defined by the user. What is obviously required is some measure of the semantic elements in a program, and for these to be used to direct the genetic operators that combine partial programs to construct new programs.

This problem with using *blind* genetic operators has been mentioned by [11], where extensions to the GP paradigm using a **DECOMPOSE** and **SPECIALIZE** operator have been suggested. They suggest a building block operator that uses a better subtree selection technique used at *"judicious system determined times"*. Unfortunately, the authors do not fully explain how these operators are used, nor how the detection of a subtree that is useful is determined.

This negative result has led us to consider the application of genetic operators in an environment that promotes the semantic analysis of program components through the field of *Inductive Logic Programming*. We will now use the relational language of horn clauses to replace the previous functional language. This is required as the field of GP is *functional*, whereas ILP is represented as a *relational*, horn clause language.

3 Inductive Logic Programming

Genetic approaches to learning combine a local search operator (mutation) and a second operator (crossover) that searches the area *between* existing individuals in the population. While these operators are highly successful in relatively simple fields such as attribute/value (ie propositional) learning, they exhibit serious limitations in more complex fields such as recursive or relational learning, as our previous example shows. Thus for example, having learnt partial programs for the membership predicate (as PROLOG statements):

```
member(X,Y) :- Y = cons(_,cons(X,_)).
member(X,Y) :- Y = cons(_,cons(_,cons(X,_))).
```

it is not at all obvious how to combine these (using crossover and mutation) to create the required recursive generalisation:

```
member(X,Y) :- Y = cons(_,Z), member(X,Z).
```

Further, in applying the genetic operators, typical genetic systems are equally likely to generalise as to further specialise a predicate that is already known to be too general. While this inefficiency has been found tolerable in simple problems, it is unlikely to be acceptable in problems with the computational complexity of recursive or relational learning. For example, a natural resource application describing **nitrate concentrations** may have a structure similar to the member construct:

```
nitrate_concentration(X,high) :- landuse(X,piggery).
nitrate_concentration(X,high) :-
                        adjacent(X,Y),
                        down_hill(X,Y),
                        nitrate_concentration(Y,high).
```

The relatively new field of inductive logic programming (ILP) has produced a plethora of interesting and demonstrably useful operators. These range from the inverse resolution operators (which construct new concepts out of a pair of existing concepts) to the Relative Least General Generalisation (RLGG) operator, which constructs a new, more general concept from a set of existing concepts [10].

Interestingly, these operators are highly non-deterministic. In any situation, there are usually numerous potential applications of inverse resolution. Inverse resolution systems have traditionally treated this as a problem to be overcome, with the system imposing artificial restrictions on the application of inverse resolution to attempt to restrict the search space [9]. Similarly, the RLGG is theoretically unique, but it is also often infinite. So RLGG-based systems compute, not the true RLGG, but an approximation to it. Again, this approximation process is highly nondeterministic, but RLGG systems typically take a fixed approach and compute a very specific approximation.

Yet, while these systems have shown an ability to learn complex and difficult programs (eg quicksort, heapsort) [9], theoretical difficulties, combined with a single-minded, deterministic learning approach have rendered them unable to learn some relatively simple real-World concepts [8]. It appears that the nondeterministic search approach characteristic of genetic learning has much to offer in such domains.

At the outermost level, the approach uses a typical genetic learning structure, the major difference lying in the operators used. In this approach, the system attempts to evolve a set of rules which, as a whole, cover the training set; it thus uses a speciation-encouraging reward mechanism [2]. However, prediction in this system will not be one stage, rather some rules may require the results of others to generate useful predictions, so a 'bucket-brigade' approach will be necessary [3].

Initially, we propose to investigate operators based on resolution and inverse resolution. The clausal form of a rule is obtained by forming the set consisting of the head of the rule, together with the negations of the items in the body of the rule. To use one of Muggleton's examples, the rule

```
heavier(A,B) :- denser(A,B), larger(A,B).
```

is represented by the clause

```
{ heavier(A,B), ~denser(A,B), ~larger(A,B) }.
```

For simplicity, the discussion below will be mostly given in terms of propositional resolution; extension to the relational case is well understood, if complex [9, 10].

3.1 Mutation Like Operators

The changes to the mutation process are relatively straightforward. Instead of a single mutation operator, mutation operators will be grouped into three classes, each of which are available only in specific circumstances. **Generalising mutations** will only be available for application when the existing concept is known (from comparison with the training data) to be too special. Generalising mutations include:

- deleting a conjunct from the concept
- replacing a term with a variable of the same type

Specialising mutations will only be available for application when the existing concept is known (from comparison with the training data) to be too general. Specialising mutations include:

- adding a conjunct of the same type to the concept
- replacing a variable with a term (of appropriate type) (guidance from the training data might be used to construct an appropriate term)

Transforming mutations will only be available for application when the existing concept is known (from comparison with the training data) to be neither too general nor too special. Transforming mutations include:

- replacing a conjunct with another
- replacing a term with another term of the same type
- replacing a variable with another variable of the same type

3.2 Crossover Like Operators

In addition to the typed crossover operator previously described, we will investigate the use of resolution and inverse resolutions operators as crossover operators.

The operators will only be applied when the two rules involved are known (from the training data) to be either both too special or both too general; in other cases, normal (typed) crossover will apply, randomly repartitioning the predicates from the two rules into two new rules (to guarantee that the resultant clauses are rules, it is required that the two clauses each contain at least one positive literal).

When the two rules are known to be too general, the resolution operator will be used. Given two clauses (see figure 1)

```
    A = {A1,...,Am,L} and B = {B1,...,Bn,~L}
resolution creates the resolvent
            D = {A1,...,Am,B1,...,Bn}.
```

That is, it conjoins the two clauses, deleting a pair of complimentary literals. When the two rules are known to be too specific, inverse resolution operators

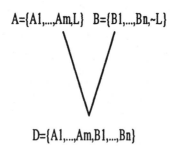

$$A=\{A1,...,Am,L\} \quad B=\{B1,...,Bn,~L\}$$

$$D=\{A1,...,Am,B1,...,Bn\}$$

Fig. 1. The V Operator

will be used. There are five of these, the two 'V' operators, which run a single resolution backwards. The absorption operator forms the most general clause. We note that more specific clauses may be created by including terms (from A1..Am) as part of the resolved clause. Using the above example, **absorption** creates B from D, whilst **identification** creates A from D and B.

```
Absorption creates B = {B1,...,Bn,~L} from
            D = {A1,...,Am,B1,...,Bn} and A = {A1,...,Am,L}.

Identification creates A = {A1,...,Am,L} from
            D = {A1,...,Am,B1,...,Bn} and B = {B1,...,Bn,~L}.
```

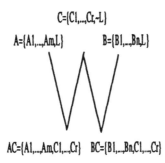

Fig. 2. The W Operator

The two 'W' operators, which run two 'adjacent' resolutions backward: given the clauses

```
    AC = {A1,...,Am, C1,...,Cr} and BC = {B1,...,Bn, C1,...Cr}
construct the clauses
    A= {A1,..Am,L}, B = {B1,...,Bn,L} and C = {C1,...,Cr,~L}.
                        (Intra-construction)
```

Inter-construction is almost the same, but replaces L by ~L. Muggleton and Buntine draw particular attention to one important aspect of the W operators: their ability to invent new predicates (the literal **L** is not present in the original data and background knowledge). Use of W operators as genetic crossovers will extend this ability to genetic programming, providing a powerful new tool for the concise expression of concepts.

The final inverse resolution operator, truncation, is effectively a degenerate case of the W operators. But, it is most readily understood as an LGG operation: from the two literals L1 and L2 truncation creates their LGG L. It will be explained in a relational context, since it does not make sense propositionally. Given two negative singleton clauses, ~L1 and ~L2, there may be a literal L which given the appropriate variable substitutions (unification), can resolve separately with each of them, giving the empty clause in each case. Running this backward, given ~L1 and ~L2, we may hope to generate such an L. In fact, a suitable choice is the Least General Generalisation (LGG) of the two, if it happens to exist.

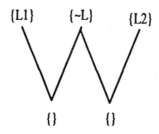

Fig. 3. The Truncation Operator

Thus from

```
        member(1,cons(1,2)).
        member(red,cons(red,green))
truncation generates:
        member(A,cons(A,B)).
```

4 Conclusion

Genetic operators are usually inappropriate for the discovery of recursive relationships. A combined approach using ILP and GP constructs seems worthy of research for building these generalised descriptions. A system that builds descriptions in an inductive manner must incorporate both a syntactic and semantic approach to construction, otherwise patterns that should be recognised *will not be found*. We have outlined a system that incorporates ILP constructs in a genetic environment, which brings together the advantages of both broad beam searching and sound logical operations for constructing generalisations.

5 Acknowledgements

The authors express their thanks to Kerry Taylor of ANU for her thoughts on using ILP constructs with genetic operators, the *Machine Learning Group, Canberra*, for thoughts, comments and encouragement, and the anonymous referees for their constructive criticism.

References

1. John Allen. *Anatomy of Lisp*. McGraw-Hill Computer Science Series, 1978.

2. Y. Davidor. A naturally occuring niche and species phenomenon — the model and first results. In R.K. Belew and L.B. Booker, editors, *Proceedings of the 4th International conference on Genetic Algorithms*, pages 257–263. Morgan Kaufmann Publishers, San Mateo, California, 1991.

3. John H. Holland. *Adaptation in Natural and Artificial Systems*. MIT Press, second edition, 1992.

4. John R. Koza. *Genetic Programming:on the programming of computers by means of natural selection*. A Bradford Book, The MIT Press, 1992.

5. John R. Koza. *Genetic Programming:on the programming of computers by means of natural selection*, chapter 19, pages 480–526. A Bradford Book, The MIT Press, 1992.

6. John R. Koza. *Genetic Programming:on the programming of computers by means of natural selection*, chapter 18, pages 473–477. A Bradford Book, The MIT Press, 1992.

7. J.R. Koza. Hierarchical genetic algorithms operating on populations of computer programs. In *Proc. 11th International Joint Conf. on Artificial Intelligence (IJCAI)*, volume 1, pages 768–774. Morgan Kaufman, San Mateo, Calif.,, 1989.

8. R.I. McKay. Relational learning for geospatial problems. Technical Report CS15/94, University College, University of New South Wales, 1994.

9. S. Muggleton and W. Buntine. Machine invention of first-order predicates by inverting resolution. In S. Muggleton, editor, *Inductive Logic Programming*, pages 261–281. Academic Press Inc. San Diego, CA 92101, 1992.

10. S. Muggleton and C. Feng. Efficient induction of logic programs. In S. Muggleton, editor, *Inductive Logic Programming*, pages 281–299. Academic Press Inc. San Diego, CA 92101, 1992.

11. U. O'Reilly and F. Oppacher. An experimental perspective on genetic programming. In R. Manner and B. Manderick, editors, *Parallel Problem Solving from Nature 2*, pages 331–340. Elsevier Science Publishers B.V., 1992.

An Application of Genetic Programming to the 4-Op Problem using Map-Trees

Tevfik Aytekin, E. Erkan Korkmaz and H. Altay Güvenir

Bilkent University, Computer Engineering and Information Science Department,
Ankara 06533 TURKEY

Abstract. In Genetic programming (GP) applications the programs are
expressed as parse trees. A node of a parse tree is an element either from
the function-set or terminal-set, and an element of a terminal set can be
used in a parse tree more than once. However, when we attempt to use
the elements in the terminal set at most once, we encounter problems in
creating the initial random population and in crossover and mutation operations. 4-Op problem is an example for such a situation. We developed
a technique called *map-trees* to overcome these anomalies. Experimental
results on 4-Op using map-trees are presented.

1 Introduction

Genetic algorithms, by combining the survival of the fittest among string structures with a randomized genetic information exchange, try to form a search
algorithm similar to the evolution process in nature. In every generation, a new
set of strings is created using bits and information coming from the fittest of the
previous generations. See [4] and [3] for details on GAs.

Genetic programming (GP) on the other hand employs programs instead of
strings [5]. Both genetic methods differ from most of the search techniques in
that they simultaneously involve a parallel search involving a large number of
points. In GP this is done by the random creation of a population of individuals
represented by programs which are the candidate solutions to the problem. These
programs are expressed in GP as parse trees. The individuals in the population
then go through a process of evolution.

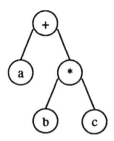

Fig. 1. A parse tree example.

Thus for example, a simple program that computes "$a + b * c$" would be expressed as in Fig.1, or to be precise as suitable data structures linked together to achieve this effect.

The programs in the population are composed of elements from a *function-set* and a *terminal-set*, which are typically fixed sets of symbols selected to be appropriate for the solution of problems in the domain of interest. The initial population consisting of individual programs is randomly created after determining these two sets. In GP the genetic information exchange is done by taking randomly selected subtrees in the individual programs and exchanging them. This is the recombination operation which is referred to as *crossover* because of the way that genetic material crosses over from one chromosome to another. Because of the closure property of the functions and terminals, this genetic crossover operation always produces syntactically legal parse trees as offspring regardless of the selection of parents or crossover points.

The crossover operation takes place in an environment where the selection of who gets to mate is a function of the *fitness* of the individual, i.e. how good the individual is at competing in its environment. Some GP techniques use a simple function of the fitness measure to select individuals (probabilistically) to undergo genetic operations such as crossover or *reproduction* (the propagation of genetic material unaltered). This is called *fitness proportionate selection. Mutation* also plays a role in this process, though it is not the dominant role that is popularly believed to be the process of evaluation, i.e. random mutation and survival of the fittest. It cannot be stressed too strongly that the GP is not a random search for a solution to a problem. The GP uses stochastic processes, but the result is distinctly better than random.

The GP executes the following cycle: Evaluate the fitness of all individuals in the population; Create a new population by performing operations such as crossover, fitness proportionate reproduction and mutation on the individuals based on the fitness; Discard the old population and iterate using the new population. One iteration of this loop is referred to as a *generation*.

As a last remark, we will state an important point that was pointed out by Koza [5]:

> Seemingly different problems for a variety of fields can be reformulated as problems of program induction (requiring the discovery of a computer program that produces some desired output when presented with particular inputs), GP paradigm provides a way to search the space of possible computer programs for an individual program that is highly fit to solve the problems of program induction.

The reason behind reformulating various problems as problems of program induction is because computer programs have the flexibility and complexity needed to express the solutions to a wide variety of problems and there is a way to solve the problem of program induction which is the GP paradigm.

Usually in GP applications there is no restriction on the number of function-set and terminal-set elements used. However in some applications there may be a restriction on the number of occurrences for each element of these sets. In this case standard crossover and mutation operation will lead to illegal parse trees. In this paper we present such an application called 4-Op where the function-set is $\{+, -, /, *\}$ and the terminal-set consists of six integers.

The next section gives a description of the 4-Op problem. Section three presents our formalism called *map-trees* which helps to redefine the crossover and mutation operations to guarantee that off-springs are legal parse trees. The fourth section makes an empirical study of our new technique and the last section concludes with an overall evaluation.

2 Description of the 4-Op Problem

4-Op is a well known TV-game where the players try to find an arithmetical expression, involving six integers, whose value is closest to a given target value. The expression may contain any number of the four arithmetical operations "$+, -, *, /$". The first four of the input number set are between one and ten and the last two are chosen from the set $\{25, 50, 75, 100\}$. The target value is between 100 and 999. An important restriction is that the players can use each element of the input set at most once.

For example, let the input number set be $\{2, 3, 5, 8, 25, 100\}$ and the target value be 467. The expression $(5 * 100) - (25 + 8) = 467$ is one of the possible answers to the question. However it is not always possible to find an exact solution. A player can get points if no other player has a closer expression.

In genetic programming applications usually there is no restriction on how many times each element of the terminal set can be used. However in our problem we can use each element at most once. So this brings a restriction to parse trees formed and to the operations on the parse trees like crossover and mutation. We can not perform crossover and mutation operations at an arbitrary point in the parse trees, since this may cause repetition of a terminal-set element in the parse tree. Let us illustrate these anomalies with an example. Consider the two parse trees named P1 and P2 in Fig.2a. The tree P1 stands for the expression $(* 5 (+ (* 3 25) 50))$ and P2 stands for the expression $(+ 3 (* (+ 5 (- 8 2)) 25))$ in prefix notation. The numbers near each node of the tree represent the crossover points. Now let us perform a crossover at points 4 on P1 and 3 on P2. The crossover fragments are shown in Fig.2a inside dashed lines. After the crossover operation, we get two offsprings as shown in Fig.2b.

Also if we consider a mutation at point 7 on P1 in Fig.3a and if we generate the mutation fragment as in Fig.3b, we get the off-spring shown in Fig.3c after the mutation operation.

Now, let us examine the trees we get after mutation and crossover. In all of them at least one element of the terminal set is used more than once. In O1 the element 5, in O2 the element 3 and in new-P1 the element 3 and 5 are used twice. Hence, the new form of expressions we have are invalid and cannot be used as solutions to our initial problem (Since there is a restriction that we can use each element of the terminal-set at most once). However this is not the case

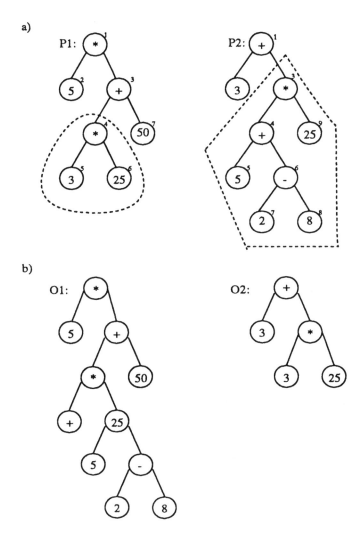

Fig. 2. a) Parents in crossover, where crossover fragments are enclosed in dashed lines. b) Offsprings after crossover. Note that both offsprings contain an element used twice.

for all crossover points and for all mutations. For instance, a crossover operation at points 2 on P1 and 4 on P2 will not violate our problem constraints.

The trees obtained after this crossover can be seen in Fig.4, and these are valid parse trees since each element of the terminal set appears at most once. Similarly we can find mutation points which generate valid parse trees.

The main problem here is to develop the appropriate data structures and techniques to overcome the illustrated anomalies. The data structures and techniques we used are not specific to our problem, but can be considered as a general approach to solving problems by using genetic programming where each element

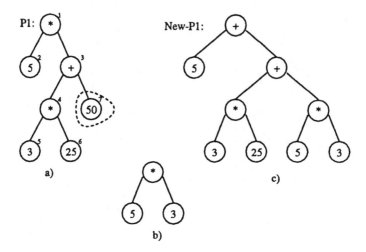

Fig. 3. a) An individual in mutation; mutation fragment is shown in dashed lines. b) Generated fragment for mutation. c) Offspring after mutation.

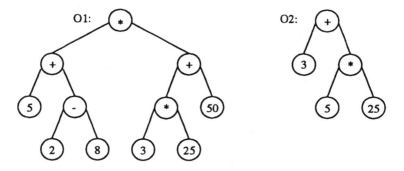

Fig. 4. Valid off-springs after crossover.

of the terminal set can be used at most once.

Before presenting our solution, to get an insight of our problem, let us analyze the search space. The search space for a GP, whose target language is LISP, is the space of all possible LISP S-expressions that can be recursively created by compositions of the available functions and available terminals for the problem.

In our problem the cardinality of the function-set is 4 and the cardinality of the terminal-set is 6. We define a valid tree as follows:

- each of the internal nodes should be an element of the function-set
- each of the leaf nodes should be an element of the terminal-set
- consists of at most 11 nodes
- leaf nodes should be distinct
- its depth should be at least 1
- every node except the leaves must have exactly 2 children, leaves do not have any children

Note that the first two of the conditions given above are from the definition of GP, and the last one is just a property of the function set used in 4-Op.

Any valid tree is a sample point in the search space. In order the find the number of points in the search space we should count all the possible valid trees that can be created. Given n terminal elements the number of different valid tree topologies that can be generated is equal to the different paranthesizations of a sequence of n numbers which is $K(n-1)$. This is known as Catalan numbers where:

$$K(n) = \left(\frac{1}{n+1} \right) C(2n, n) \qquad (1)$$

Here, C denotes the combination operation.

The valid trees we can generate will have at least 2 and at most 6 children. We will divide our computation into classes where $class(n)$ contains the set of trees with exactly n leaves. After determining all different valid tree topologies in each class, we are going to compute the number of different valid trees we can create using the given function-set and terminal-set. For an illustration consider the valid tree topology shown in Fig.5. The numbers in each node represents the number of different choices we can insert into that node. Since there is no restriction on the choices of the functions as terminals at each internal node we have 4 choices. However, since we cannot use a terminal-set element more than once, at each external (leaf) node we have a decreasing sequence of choices. Therefore, in for the valid tree topology shown in Fig.5, there are $*4*4*6*5* 4*3 = 23040$ different valid trees.

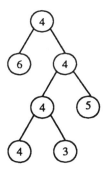

Fig. 5. A valid tree topology. The numbers represent the number of different choices for a node.

Let us now compute the number of all valid trees we can create given a specific terminal-set and function-set whose cardinalities are 6 and 4, respectively.

$class(2)$: $K(1) = 1$ $1 * (4 * 6 * 5) = 120$
$class(3)$: $K(2) = 2$ $2 * (4 * 4 * 6 * 5 * 4) = 2608$
$class(4)$: $K(3) = 5$ $5 * (4 * 4 * 4 * 6 * 5 * 4 * 3) = 115200$
$class(5)$: $K(4) = 14$ $14 * (4 * 4 * 4 * 4 * 6 * 5 * 4 * 3 * 2) = 2580480$
$class(6)$: $K(5) = 42$ $42 * (4 * 4 * 4 * 4 * 4 * 6 * 5 * 4 * 3 * 2 * 1) = 30965760$

For example, in $class(4)$ there are 5 different tree topologies, and 115200 different valid trees. Therefore, we can create a total of 33,666,168 different valid trees, i.e we have 33,666,168 sample points in the search space.

3 Solution

The easiest way to handle the anomalies discussed in the previous section is to generate crossover and mutation operations as usual and then discard the invalid parse trees. However when we implemented this solution, we saw that it is a very inefficient way to handle our problem, because about half of the population were formed with such invalid parse trees and we had to discard all of them.

Random keys, developed by Bean and Norman could be another solution [1, 2]. Random keys are developed to overcome the difficulty of genetic algorithms maintaining feasibility from parent to off-spring. To illustrate the use of random keys, consider a simple genetic algorithm approach to the traveling salesman problem. A candidate solution to a TSP is a tour through n cities. Two such tours for a map of five cities are 2-1-3-5-4 and 4-2-3-1-5. Consider a crossover operation after the second city, then resulting off-springs are 4-2-3-5-4 and 2-1-3-1-5. Neither of these is a valid tour. As it can be seen in TSP a city cannot occur in a solution more than once, at first glance we may think that this is exactly the same problem we have in 4-Op, so that we can use random keys to overcome anomalies described in section two. However what makes our problem different is that, in GAs the strings have constant lengths but in GP the parse trees have variable sizes. This difference causes improper probabilistic distribution of terminal-set elements and we may have repetition of keys in later generations.

However we were able to develop another technique and a suitable data structure to overcome this problem. Before explaining our solution let us define some notions.

The function $S(T, node)$: returns the set of terminal-set elements appearing at the leaves of the tree rooted at node whose infix order numbering is $node$ in a tree T. Fig.6 gives the values of this function on an example tree.

We can state the necessary condition to guarantee having valid off-springs after crossover and mutation operations. Let $T1$ and $T2$ be two parse trees. An off-spring obtained by crossover operation applied to x of $T1$ and y of $T2$ is a valid tree if

$$(S(T1, 1) - S(T1, x)) \cap S(T2, y) = \emptyset \tag{2}$$

A crossover operation using map-trees is shown in Fig.7. In this example, the crossover points are 4 on T1 and 6 on T2.

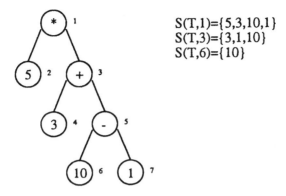

$S(T,1)=\{5,3,10,1\}$
$S(T,3)=\{3,1,10\}$
$S(T,6)=\{10\}$

Fig. 6. Values of the function $S(T, node)$ on an example tree.

$$(S(T1, 1) - S(T1, 4)) \cap S(T2, 6) = \{\{5, 3, 25, 50\} - \{3, 25\}\} \cap \{2, 8\}$$
$$= \{5, 50\} \cap \{2, 8\}$$
$$= \emptyset.$$

Therefore, Off-T1 is a valid tree. However, since

$$(S(T2, 1) - S(T2, 6)) \cap S(T1, 4) = \{\{3, 5, 8, 2, 25\} - \{2, 8\}\} \cap \{3, 25\}$$
$$= \{3, 5, 25\} \cap \{3, 25\} = \{3, 25\}$$
$$\neq \emptyset,$$

Off-T2 is not a valid tree.

Also let $S(Tm, z)$ be the set of terminal elements of the mutation subtree and x be the mutation point on $T1$. The resulting off-spring is a valid tree if

$$(S(T1, 1) - S(T1, x)) \cap S(Tm, z) = \emptyset \tag{3}$$

In our implementation, we first check if the crossover points are valid for parents $T1$ and $T2$. If they are not valid for both of them we generate randomly two other crossover points and continue the process until we can generate a valid offspring at least for one of the trees. If the crossover is valid for only one of the trees then we generate the valid offspring and reproduce the remaining tree.

It is not possible to implement the set operations using only the parse trees because of the time efficiency reasons, so we have used another data-structure. For each parse tree, we also store the *map-tree*. A node of a map-tree stores the set of terminal-set elements occuring in the leaves of the subtree rooted in that node. A parse tree and its corresponding map-tree are shown in Fig.8.

The set operations are carried out on this tree more efficiently. It can be easily seen that the map tree can be constructed by exchanging every node of the parse tree with the set returned by $S(T, node)$.

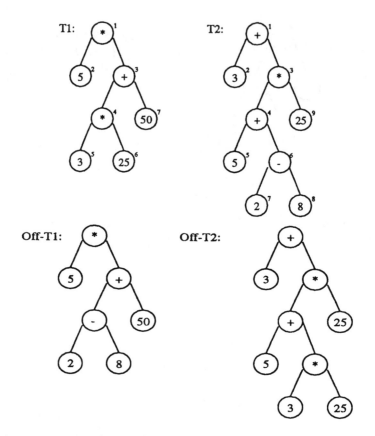

Fig. 7. A crossover operation using "map-tree." Crossover points are 4 on T1 and 6 on T2.

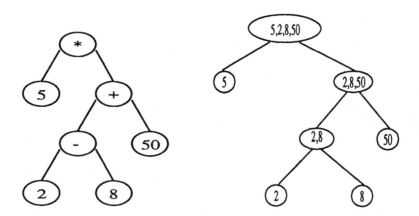

Fig. 8. A parse tree, and its corresponding map-tree.

Other than crossover and mutation anomalies, creating the initial random population is another problem that we have encountered. Since the individuals in the initial population are created randomly, this may easily lead to forming illicit parse trees where there is a repetition of terminal-set elements. Preventing such parse trees in the initial population is easier than preventing mutation and crossover anomalies. The idea is that, for each individual in the population, after choosing a terminal-set element randomly, this element is discarded from the set so that repetition of elements is prevented.

4 Empirical Evaluation

We have used 110 randomly generated input data in order to test and examine the results of our technique. In these experiments the population size used is 250, mutation rate is 10% and number of generations examined is 20. 10% mutation rate can be considered rather high since in genetic programming applications the mutation rate is usually zero. However the search space of our problem is relatively small and the loss of genetic information due to randomness of mutation can be recovered by crossover operations. On the other hand 10% mutation rate in this problem provides a means for recovering from local maximas and leads to a better examination of the search space.

In Table 1, for the following five input data sets, the values of the average fitness and best fitness versus generation numbers are given. In these five selected examples we can have an insight of how our program approaches to the target value in each generation. The data sets are:

Data set 1: Input integers are {2,4,6,7,25,75} and target value is 458
Data set 2: Input integers are {1,3,5,9,25,50} and target value is 846
Data set 3: Input integers are {1,4,8,9,25,50} and target value is 359
Data set 4: Input integers are {4,6,7,9,25,75} and target value is 793
Data set 5: Input integers are {2,3,7,9,25,100} and target value is 458

In Table 2 the test results for the 110 random input data sets are grouped according to fitness measure. As it can be seen in Table 2, in 40% of the test results we have found an exact solution. If we consider that some input data sets do not contain exact solutions, we can claim that these test results are successful.

The graphs given in Fig.9 and Fig.10 show the average of "average fitness" values versus generation number and average of "best fitness" values versus generation number. In these figures the fitness of a tree is computed as the absolute value of the difference between the target value and the value of the expression represented by the given tree. As it can be seen in Fig.9 after the dramatic fall in the first generation, although there is a fluctuation due to the high mutation rate (10%), the average of "average fitness" of the population shows a decreasing behavior throughout the generations. In Fig.10 the average of "best fitness" values decreases steadily, and after nineteen generations the value of the average of "best fitness" reaches to 1.5.

Table 1. Average and best fitness values versus generation.

	Dataset 1		Dataset 2		Dataset 3		Dataset 4		Dataset 5	
Gen	Avg.	Best	Avg.	Best	Avg.	Best	Avg.	Best	Avg.	Best
0	2250.8	11	2746.5	46	4027.4	9	4446.7	14	5210.4	8
1	479.11	4	1266.8	21	339.2	9	1833.4	7	996.5	8
2	350.4	4	867.8	21	996.1	9	514.1	7	2849.0	8
3	260.0	2	808.8	4	271.2	9	373.8	7	495.3	8
4	206.3	2	473.6	4	282.7	9	426.8	7	5554.6	4
5	263.6	2	455.5	4	418.3	9	309.5	7	2326.3	4
6	672.7	2	501.2	4	477.0	1	292.1	0	685.5	4
7	1043.9	1	839.3	4	320.2	1	-	-	1196.8	3
8	667.4	1	398.7	4	2094.9	1	-	-	1426.0	3
9	174.7	1	365.1	4	276.6	1	-	-	3311.4	3
10	123.2	1	233.3	4	206.1	1	-	-	2225.6	3
11	802.2	1	655.7	4	160.9	1	-	-	1275.7	3
12	255.6	1	510.1	4	199.1	1	-	-	842.6	3
13	126.0	0	1004.9	4	199.1	1	-	-	905.1	3
14	-	-	537.0	4	190.9	1	-	-	1065.7	3
15	-	-	431.9	4	157.5	1	-	-	1536.2	3
16	-	-	425.3	4	244.1	1	-	-	152.3	3
17	-	-	451.1	4	185.3	1	-	-	191.4	3
18	-	-	537.6	4	135.0	1	-	-	157.6	3
19	-	-	1241.1	4	283.7	1	-	-	255.5	3

Table 2. Fitness measures by grouping.

Fitness Measure	Number of Times
0	44
1	31
2	12
3	7
4	4
5	6
6	0
7	0
8	0
9	1
10	1

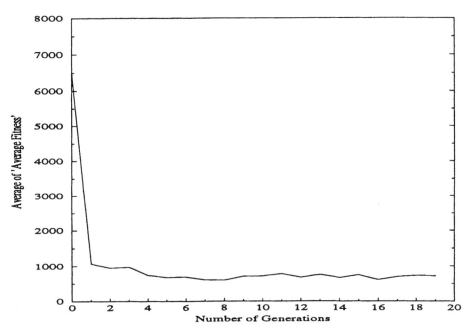

Fig. 9. Average of "average fitness" versus number of generations.

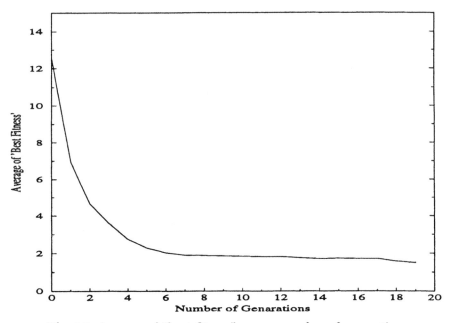

Fig. 10. Average of "best fitness" versus number of generations.

5 Conclusion and Future Directions

The genetic programming paradigm provides a way to solve a wide variety of different problems from many different fields. These problems can be reformulated as the problems of program induction. When we have a problem whose search space is well characterized and if we have also a good heuristic to solve the problem possibly genetic programming would not give a better result. However it is very convenient to use genetic programming when we do not know how to approach to the problem. Various applications of GP on many different subjects provide considerable evidence of the generality of the GP paradigm.

In standard GP the user determines the elements in the function set and the terminal set. But he/she can not put a restriction on the number of times of their usage, i.e. on the number of times of the occurrences in the parse trees. Restrictions on some problems make standard GP inapplicable. 4-Op problem is one of them and it puts a restriction on the number of times of using the terminal-set elements. More specifically a terminal-set element can be used at most once. Our technique makes GP applicable to 4-Op problem. In the experimental results we have observed that our program has given 40% exact solutions and after about eight generations the average of "best fitness" is below two. These results indicate that our technique is effective in the solution of this kind of problems.

Our technique is not specific to 4-Op problem. It can be extended, without changing the idea behind it, to problems that limit the use of not only terminal-set elements but also function-set elements. It can be used for all problems where the elements of the terminal or the function-set are to be used for a specific number of times.

References

1. Bean J.C.: Genetic and Random Keys for Sequencing and Optimization. Dept. of Industrial & Operations Engineering, Univ. of Michigan, Technical Report (June 1992) 92-43
2. Bean J.C. and Norman B.: Random Keys for Job Scheduling. Tech. Report, Dept. of Industrial and Operations Engineering, Univ. of Michigan, Ann Arbor (January 1993)
3. Goldberg D.E.: Genetic algorithms in search, optimization, and machine learning. Addison-Wesley (1989)
4. Holland J.H.: Adaptation in natural and artificial systems. University of Michigan Press (1975)
5. Koza J.R.: Genetic programming on the programming by means of natural selection. Cambridge, MA: The MIT Press (1992)

Direct Replacement: A Genetic Algorithm Without Mutation Which Avoids Deception[1]

Jon Rowe and Ian East
Department of Computer Science
University of Buckingham
Buckingham
England MK18 1EG

ABSTRACT

A Genetic Algorithm is introduced in which parents are replaced by their offspring. This ensures there is no loss of alleles in the population, and hence mutation is unnecessary. Moreover, the preservation of less fit alleles in some members of the population allows the GA to avoid falling into deceptive traps.

1   INTRODUCTION

A Genetic Algorithm can run into trouble when certain alleles, not belonging to the optimal solution, lead it astray by offering above average fitness values. Such alleles will tend to swamp the population before the alternatives have had a chance, in accordance with the Schema Theorem (Holland, 1975). Traditionally, the mutation operator has been used to try to prevent allele loss. We show that, for reasonable mutation rates, this is futile. Other methods include manipulating the fitness function (for example, in scaling and sharing (Goldberg, 1989)) and changing the representation (as in "messy GAs" (Goldberg, Deb, Kargupta, & Harik, 1993)). We propose a technique called **direct replacement** which guarantees no allele loss. It is shown to find the optimal solution to three deceptive test functions, where a standard GA will quickly fall into the trap.

2   DIRECT REPLACEMENT WITH NO MUTATION

One problem that can lead a GA into an early convergence at a false optimum is loss of alleles. Mutation is usually used to try to overcome this. However, mutation is of little use when there is selection pressure in favour of a deceptive schema, as it would need to simultaneously mutate all genes in such a schema to the "correct" (that is, optimum) values.

---

[1] This work was funded under SERC grant number GR/J 14301

For example, if the optimal solution contained the schema 00000***** but there is selective pressure in favour of 11111*****, then mutation would need to change all five incorrect genes in the same go. Obviously, if this was known in advance, the representation scheme could be changed to accommodate the problem. However, in real problems we can rarely predict where the problems will lie.

If the deceptive schema is of order a, and the probability of a gene mutating is p, then the probability of escaping deception is p^a. The number of attempts expected to reach the solution is:

$$\frac{1}{p^a}$$

If we express the order as a fraction of the chromosome length, n, as:

$$a = \lambda n$$

then the expected number of attempts is bigger than the search space size if and only if:

$$\frac{1}{p^{\lambda n}} > 2^n$$

that is:

$$p < 2^{-\frac{1}{\lambda}}$$

One recommended mutation rate is $\frac{1}{n}$, (Back, 1993). At this rate, the above condition becomes:

$$\lambda > \frac{1}{\log_2 n}$$

So, for example, if $n > 32$, it would be quicker to enumerate the search space than to rely on mutation to find a way out of a trap that is 20% of the chromosome length. The similar results apply for any typical mutation rate, independent of n (e.g. 0.01). (For a more thorough theoretical treatment of the issues, see (Spears, 1992)).

If, then, we cannot rely on mutation to supplement the allele population in such a way as to avoid deception, another method must be sought. One possibility is to circumvent the loss of any alleles (this happens, in a sense, in order-based GAs, but the generalisation of the notion of an "allele" requires care (Radcliffe, 1991)). If, when two chromosomes crossover, they are both replaced by the resulting offspring, then all the original alleles are preserved. This is called direct replacement.

A steady-state GA with direct replacement and no mutation works as follows:

1. Generate a random population

2. Choose two chromosomes from the population with probabilities proportional to their fitness. These are the parents.

3. Use a one point crossover on these to create two **offspring**.

4. If **either** of the offspring is fitter than **both** parents, then replace both parents with **both** offspring.

5. If either offspring is the optimum solution then stop.

6. Go back to step 2.

What size population is needed to ensure that all alleles are present? Consider, to start with, one gene position. Suppose there are k alleles for this position. In a population of size M the probability that all alleles are present is:

$$\frac{k!\,S(M,k)}{k^M}$$

where $S(M,k)$ is the Stirling number of the second kind. This formula comes from counting the number of onto functions from M population places to k possible alleles, divided by the total number of possible functions.

If all gene positions have the same number of alleles, then the probability that all alleles for every gene position are present is:

$$\left(\frac{k!\,S(M,k)}{k^M}\right)^n$$

The following table (Table 1) gives the smallest population sizes that make this probability at least 99.9% for various chromosome lengths and cardinalities.

k	n=50	n=100	n=150	n=200
2	17	18	19	19
3	30	32	33	33
4	43	45	47	48
5	56	59	61	62
6	70	73	76	77
7	83	88	90	92
8	97	102	105	107
9	111	117	120	123
10	125	132	135	138

Table 1: Minimum population size that gives all possible alleles with probability greater than 99.9%

3   EVADING AN OUTLAW GENE

The working of the direct replacement GA can be illustrated with an "outlaw" function. In a chromosome, one particular gene is distinguished as the outlaw. If this gene has value 0, then the fitness of the chromosome is the total number of zeros in it (including the outlaw zero). If, however, the outlaw gene has value 1 then the fitness is a fixed amount, close to, but just less than, the chromosome length, which is the optimum. The outlaw gene tries to tempt the GA with an offer of "get fit quick."

A standard GA will rapidly select for the outlaw gene with value 1. In any chromosome containing this value, there is no selective pressure on the rest of the chromosome. Hence, even if mutation switches the outlaw off, there is little chance that there will be enough zeros to hold it there. In a number of tests with a standard GA (described in more detail in the next section), with a chromosome length (and hence optimum) of 20 and an outlaw value of 19, the outlaw gene swamped the population in less than 100 generations (a generation being the product of a single crossover).

Now consider what happens with the direct replacement GA. Approximately half of the population will have a zero in the outlaw position. Whenever these are involved in a crossover, this zero is preserved. There is therefore plenty of time for selection pressure to favour large numbers of zeros to collect in such chromosomes. Eventually, the fitness of these overtakes the fitness of those with the outlaw and the optimum solution is reached. In tests, a direct replacement GA (using the same function as before) with population size 100 reached the optimum after 7500 generations on average (that is, after evaluating less than 1% of the search space — remember that each generation is two evaluations) with a standard deviation of 2500.

4   COMPARISON WITH A STANDARD GA

Two, more difficult, tests were also used, based on functions from (Schaffer & Eshelman, 1991) In these the chromosome length is 50 and the population size is always 100 (in both the direct replacement and standard GA). The standard GA is steady-state, parents being chosen according to fitness, with replacement of the least fit member of the population by one offspring, chosen at random from the two (a variant of Holland's R1 algorithm (Holland, 1975)). A mutation rate of 0.01 was used. Each run was done five times and an average taken.

4.1   LOTS OF SMALL TRAPS

In this test function, a chromosome is evaluated as follows. Divide the chromosome into ten blocks of five. In each block, score five points for all zeros, otherwise score one point for two ones and two points for four or more ones. Add up the scores from each block to give the total. The optimum is therefore 50, with all zeros, but there is selective pressure towards ones.

The standard GA tended to select for all ones, with an occasional block of zeros. The best managed on one run was a score of 30. The program had always converged on its sub-optimal solution by generation 2000.

The direct replacement GA makes progress a little slower than the standard GA. Figure 1 shows the best scores found (averaged over the five runs) for the first 2000 generations. However, unlike the standard GA, the direct replacement GA can't converge, so it keeps going, reaching the optimal solution after an average of 12500 generations (note that this is less than 10^{-10} of the search space), standard deviation 1500.

4.2   A FEW BIG TRAPS

The next test function is similar to the previous one. This time, the chromosome is taken in five blocks of ten. A score of ten is given for all zeros, otherwise the score is half the number of ones in the block. Add up the scores from each block to get the total. Again the optimum is 50 with all zeros, but with even stronger selection pressure towards ones.

Tests on each algorithm were again run five times each and averaged. The standard GA always converged to a false optimum before generation 4000. As with the last test, the direct replacement GA improves at a slower rate, but eventually overtakes the standard GA and arrives at the optimum solution, after 43730 generations on average, standard deviation 15000.

Figure 2 shows the best solution found for the first 4000 generations of each algorithm. The direct replacement GA overtakes the standard GA at generation 2400.

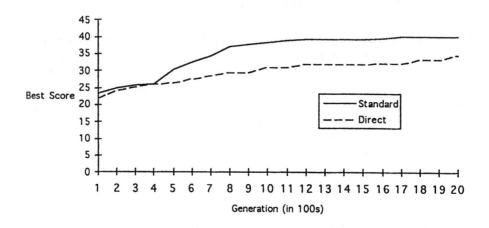

Figure 1

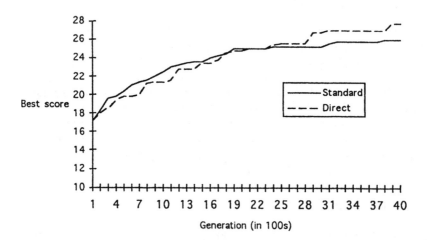

Figure 2

5 THE GENE INVARIANT GA

After completing this work, we discovered the Gene Invariant GA of (Culberson, 1993). This algorithm also avoids mutation, but uses the provision of complementary bit strings in the population to maintain allele diversity. Culberson's results reinforce our view that mutation is at best a background operation and that crossover provides the main source of power in a GA's search.

In a sense our work generalises that of Culberson, showing the power of crossover without mutation in *any* population, not just in Culberson's special population of complementary strings. Direct replacement maintains the diversity of the alleles in the population sufficiently well without the need to use complementary strings.

6 CONCLUSIONS

The direct replacement GA seems to be able to avoid deceptive traps without the aid of mutation. The way it does this bears some analogy with speciation. Since any allele will remain in the gene pool, it will become linked with those with which it can best increase fitness. Suppose a species of moth is beginning to vary. Some have a gene for, say, dark wings, while others have light wings. It is not clear which of these will become the better strategy, but while they both remain in the gene pool, they will tend to accumulate with other genes that best suit them. Hence the dark wing gene will tend to be found with a gene for settling on dark backgrounds. Similarly, the light wing gene will tend to be linked to the gene for settling on light backgrounds. Every allele will tend to do the best it can (in terms of the genes with which it associates) to survive.

It is interesting to note that, when we say "the gene for dark wings" we do not suppose that there is one position in the DNA that codes for dark wings. Rather, we refer to the total genetic basis for dark wings. From this it follows that a "mutation" giving rise to dark wings may arise from the action of crossover, bringing together the appropriate genetic material for the first time, rather than a physical mutation of the DNA. To some extent, therefore, it is possible to explain the adaptive evolution of traits in terms of crossover alone, without referring to physical mutation.

6   REFERENCES

Back, T. (1993). Optimal Mutation Rates in Genetic Search. In S. Forrest (Ed.), Fifth International Conference on Genetic Algorithms, (pp. 2-8). University of Illinois at Urbana-Champaign: Morgan Kaufmann.

Culberson, J. C. (1993). Crossover versus Mutation: Fueling the Debate: TGA versus GIGA. In S. Forrest (Ed.), Fifth International Conference on Genetic Algorithms, (p. 632). University of Illinois at Urbana-Champaign: Morgan Kauffman Publishers, Inc.

Goldberg, D. E. (1989). Genetic Algorithms in Search, Optimization and Machine Learning. Addison-Wesley.

Goldberg, D. E., Deb, K., Kargupta, H., & Harik, G. (1993). Rapid, Accurate Optimization of Difficult Problems Using Fast Messy Genetic Algorithms. In S. Forrest (Ed.), Fifth International Conference on Genetic Algorithms, (pp. 56-64). University of Illinois at Urbana-Champaign: Morgan Kaufmann.

Holland, J. H. (1975). Adaptation in Natural and Artificial Systems. Ann Arbor: University of Michigan Press.

Radcliffe, N. (1991). Forma Analysis and Random Respectful Recombination. In R. K. Belew & L. B. Booker (Eds.), Fourth International Conference on Genetic Algorithms, (pp. 222-229). University of California, San Diego: Morgan Kaufmann.

Schaffer, J. D., & Eshelman, L. J. (1991). On Crossover as an Evolutionary Viable Strategy. In R. K. Belew & L. B. Booker (Eds.), Fourth International Conference on Genetic Algorithms, (pp. 61-68). University of California, San Diego: Morgan Kauffman.

Spears, W. M. (1992). Crossover or Mutation? In L. D. Whitley (Ed.), Foundations of Genetic Algorithms 2, (pp. 221-237). Vail, Colorado: Morgan Kaufmann.

Competitive evolution:
A Natural Approach to Operator Selection

Q. Tuan Pham

School of Chemical Engineering and Industrial Chemistry
University of New South Wales, Sydney, Australia

Abstract.   One of the main problems in applying evolutionary optimisation methods is the choice of operators and parameter values. This paper propose a competitive evolution method, in which several subpopulations are allowed to compete for computer time. The population with the fittest members, and that with the highest improvement rate in the recent past, are rewarded.

When using identical strategies in the subpopulations, this competitive strategy provides an insurance against unlucky runs while extracting only an insignificant cost in terms of extra function evaluations. When using different strategies in the subpopulations, it ensures that the best strategies are used and again the extra cost is not great. Competitive evolution is at its best when an operator - or the lack of it - may have a very detrimental effect which is not known in advance.   Occasional mixing of the best performing subpopulations leads to further improvement.

Symbols

d	normalised distance between two vectors x (equation 2)
$f(x)$	function to be optimised
m	mutation frequency
m_0	base mutation frequency
N	population size
N_{eval}	number of fitness evaluations
N_{var}	number of search variables
p_{ux}	uniform crossovers / total crossovers
R	weighting factor indicating bias of offspring towards fitter parent
S	standard deviation of offspring from expected value, as a fraction of δ
x	$= (x_1, x_2...)$, vector of search variables
$x_{i,min}$	minimum allowable value of x_i
$x_{i,max}$	maximum allowable value of x_i
δ	distance for which the exponential part of the mutation frequency is 0.5

1 Introduction

Evolutionary optimisation methods such as genetic algorithms (Holland, 1975; Goldberg, 1989) and evolution programs (Michalewicz, 1992; Fogel and Atmar, 1992) start with a population of several trial points chosen at random, and let the population evolve towards the optimum using a basket of rules governing reproduction, mutation and selection. Many rules and operators for transforming the population have been proposed, but their selection is an art more than a science, and often depends as much on the user's aesthetic preferences as on the demands of the problem. This is particularly the case when little is known about the nature of the problem that would help determine the best approach. Meanwhile, the proliferation of operators (e.g. single/ double/ uniform/ multi/ arithmetic crossover, binary mutation, Gaussian mutation) and arbitrary numerical parameters (population size, mutation and crossover probabilities, etc.) make it more and more imperative to find some rational selection method. Evolutionary computation cannot become a mainstream technique, useable by non-specialists, until such guidelines are provided.

Suggestions for a logical approach to the selection of operators and parameters have been few, as shown in a review by Davis (1991, p.92). A manual or "brute force" approach has sometimes been followed, for example by De Jong (1975) and Schaffer et al (1989) who tested various combinations of population size, mutation rates, crossover probability, elitism/non-elitism and other parameters. The drawbacks of manual optimisation are obvious (indeed they are the very reason why automatic optimisation methods such as G.A. were invented!).

Grefenstette (1986) used a meta-genetic algorithm to optimise genetic algorithm; this approach succeeded because genetic algorithms (and their relative, evolution programs) are adept at handle large noise. However, even meta-optimisation methods are very time consuming (they typically involve thousands of first-level G.A.s, each of which often consist of thousands of calculations) and the results are not easily generalisable.

Davis (1989, 1991) proposed adaptive methods, where the relative merit of various operators (creep, mutation, crossover, etc.) is evaluated, depending on the frequency with which they produce an improvement. A somewhat complicated bookkeeping exercise is carried out where part of the "credit" gained by an offspring is passed back to its parents and ancestors. Every now and then the frequencies of the operators are altered according to their performance. The idea that operators, like members of a population, also evolve is appealing, but the proposed mechanism - some godlike bookkeeping superbeing controlling the process - may be unattractive to evolutionary purists. Furthermore several additional parameters are introduced, such as how often to recalculate the operators' frequency, how to assign the credit, how much of the credit is passed back to the parents, etc., and there is little intuitive guideline for these parameters.

This paper proposes a method for dealing with the problem of which operator (or parameter value) to use, in cases where the best choice is unknown. I believe the method is intuitively satisfying and aesthetically pleasing in that it does not assume a supervisory superbeing, but complies with the Darwinian evolutionary paradigm that is dear to all G.A. and E.P. practitioners.

2. Theory

2.1 Proposed Method

A football coach must make use not only of his most experienced veterans, but also of any promising up-and-coming player. His long-term success depends on striking a good balance between these categories. Similarly, nature provides a variety of mechanisms (such as cuddliness) to ensure the survival of healthy growing cubs even though these may one day threaten the dominant animals. It is with such examples in mind that the following method is suggested.

In the proposed method, which may be termed competitive evolution, several populations which may differ in evolution strategies (size, mutation frequency, crossver mechanisms, etc.) are allowed to evolve simultaneously. At each stage, the populations' performances are compared, and only the best populations are allowed to evolve further. What do we mean by the "best" populations? They are: firstly, that with the fittest member; secondly, that which has demonstrated the fastest improvement in recent cycles. (In this paper, this is equated with the population that has stalled (produced no fitter offspring than its present best) for the least number of cycles; however, other definitions are possible depending on the base algorithm used).

Each of these two are allowed to evolve for a few more steps, then another comparison is made between all the populations.

Thus, a pseudocode layout for the method is:
{
    Initiate all N_p populations
    repeat
        Find population with the fittest member
        Let this population evolve for m steps
        Find population which has stalled the least
        Let this population evolve for n steps
    until end condition satisfied
}

where m, n are parameters of the method. A preliminary test showed that the method is not very sensitive to these parameters as long as they are small, so unless otherwise stated, we will let m=n=5.

2.2 Base Algorithm

The method described above can obviously be "grafted" on any evolutionary algorithm (G.A., E.S. or E.P.). In this paper, it will be tested on an algorithm designed by the author for numerical optimisation.

Evolutionary computation has for historical reasons been classified into geneteic algorithms (G.A.), Evolutionary Strategies (E.S.) and Evolution Programs (E.P.). In broad terms, each type is associated with a different type of genetic representation (binary for G.A., floating points for E.S. and arbitrary data structures for E.P.) and a different set of operators (binary mutation and crossovers for G.A., arithmetic mutation and crossover for E.S. and problem-dependent crossover for E.P.). Selection methods also differ, with selection occurring before reproduction in G.A. and after reproduction in E.S. However, as remarked by Michalewicz (1992), in the last few years the different streams of evolutionary computation have converged more and more and make use of each other's findings. The base algorithm used by the author is probably closest to evolution strategies because it uses floating point representation and pre-reproductive selection. In G.A. parlance it employs elitism, steady-state population replacement, generational overlap (generation gap = 0), random tournament selection and an extended version of arithmetic crossover (Janikow and Michalewicz, 1991; Michalewicz, 1992).

Suppose that the value of a vector x is to be found to maximise a function $f(x)$, where x_i (i=1 to N_{var}) are real-valued variables (integer variables may be derived by the simple expedient of truncating or rounding a real variable). The search is to be carried out in the range $(x_{1,min}, x_{1,max})$, $(x_{2,min}, x_{2,max})$ etc.

The proposed evolutionary algorithm is as follows:

```
{
    Create N vectors x¹, x², ...xᴺ
    Repeat {
        Reproduction-mutation
        Selection
    } until end conditions are met
}
```

The reproduction-mutation operation introduces ONE new member, x^{N+1}, by "crossing" two existing members x^A, x^B. The selection operation reduces the population size back to N by eliminating a member.

The reproduction-mutation operation is carried out as follows (x^A denotes the "fitter" or more optimal member of the pair):

{

    Choose at random two members x^A and x^B from present population.
    Decide whether to mutate.
    for $i=1$ to N_{var} {
        if decided to mutate
                then { $\Delta_i = (x_{i,max} - x_{i,min})/4$ }
                else { $\Delta_i = x^A_i - x^B_i$ }
        $x^{N+1}_i = x^A_i + Gauss(R,S)\,\Delta_i$
        if $(x^{N+1}_i < x_{i,max})$ then { $x^{N+1}_i = x_{i,max}$ }
        if $(x^{N+1}_i > x_{i,min})$ then { $x^{N+1}_i = x_{i,min}$ }
    }

}

and the selection operation is

{

    Choose at random two members from the present population.
    Eliminate the less fit member.

}

In the above scheme, $Gauss(R,S)$ is a random normal variable with mean R and standard deviation S. R (the reproductive bias) indicates the extent to which the offspring is biased towards the fitter parent: $R=1$ means that the expected value of the offspring is the same as that of the fitter parent, x^A, while $R=0$ means that it will be the same as the less fit parent, x^B. $R=0.5$ indicates zero preference. S measures the spread of the random deviations of offspring from their expected values, as a fraction of Δ_i, and so will be called the fuzziness factor. Δ_i is the difference in x_i between the two parents, or, if a mutation occurs, a quarter of the search range for x_i.

The mutation frequency for each trait, m, is calculated from

$$m = m_0 + 2^{-d/\delta} \tag{1}$$

where

$$d^2 = \sum_{i=1}^{N_{var}} \left(\frac{x^A_i - x^B_i}{x_{i,max} - x_{i,min}} \right)^2 \tag{2}$$

The mutation frequency consists of a term m_0 which is the same for all reproductive operations and a term $2^{-d/\delta}$ which varies exponentially with the "distance" d between the two parents (as a fraction of the total search range). The idea is to keep the mutation rate low until the whole population is crowded together and diversity has been lost. Then the above expression will ensure that the rate of mutation will "explode" and re-introduce diversity.

The algorithm described above has some features which make it particularly suitable for competitive evolution: the elitist feature (retaining the best member) makes it easy to measure rates of progress, free from the effect of random fluctuations; and the steady-state strategy means that control can pass quickly from one population to another with very little inertia. Note, however, that most algorithms used these days employ elitism, which is required to ensure convergence (Rudolph, 1994).

3. Experimental Results and Discussion

3.1 Testing and optimisation of the base algorithm

Only a brief description of experiments with the base algorithm will be presented, since it is not the primary objective of this paper. Following Grefenstette's (1986) approach, an evolutionary meta-search was used to optimise the parameters for the optimisation of each of five test functions first assembled together by De Jong (1975). To avoid confusion due to two evolutionary algorithms running simultaneously at different levels, the search for the optimum of a test function (such as Rosenbrock's) will be called the inner search and each inner function evaluation an inner sample. The meta-search seeks to optimise a meta-fitness function f^{META}, defined by

$$f^{META} = Mean \ (log(N_{eval})) + 2 \ Std \ (log(N_{eval}))$$

where N_{eval} is the number of inner samples per inner search, $Mean$ the arithmetic mean, and Std the standard deviation calculated over three inner searches. f^{META} is thus a very rough estimate of the maximum number of function evaluations needed to reach the optimum.

After a few preliminary runs to estimate the search ranges, the following search ranges for the meta-search was chosen: $N = 2$ to 20, $R = 0.5$ to 2, $S = 0.02$ to 1, $log_{10}(m_0) = -5$ to -1, $log_{10}(\delta) = -5$ to -1. The meta-search successfully determined the optimal parameters for each of De Jong's five test function, although up to 3000 meta-samples may be required in some cases.

For Rosenbrock's function, the following "optimum" was obtained: $N_{opt} = 12$, $log_{10}(m_{0,opt}) = -5$, $log_{10}(\delta_{opt}) = -3.1$, $R_{opt} = 1.47$, $S_{opt} = 0.055$. Using these parameters, each inner search required on average between 300 and 360 inner samples. Notice the relatively small optimal population size N. Conventionally, population sizes of about $N = 50$ upwards have been more typical (Goldberg, 1989). The value for m_0 is right at the lower limit of the search range, indicating a "best" m_0 of 0 (since the probability of a mutation to occurs in 350 or so reproductions, due to m_0 alone, is far less than 1). On the other hand, a δ-value of $10^{-3.1}$ will still ensure a large number of mutations in the later stages of the search, when the population is crowded in a small area. It is believed that the exponential behaviour of the term $2^{-d/\delta}$ in the mutation frequency is what enables the present algorithm to work well with relatively small populations, as it ensures that diversity is recreated as soon as it is lost.

A search for the optimal parameters for De Jong's (1975) F_1 function, a three-dimensional quadratic, yielded the following values: $N = 9$, $log_{10}(m_0)$ = -4.5, $log_{10}(\delta)$ = -5, $R = 1.3$, $S = 0.7$. Again the optimal population size is much smaller than usual. In fact, $N = 2$ still manages to converge to the optimal of F_1 in reasonable time.

De Jong's F_3 function is a uniformly increasing stepping function in five dimensions. A meta-optimisation yielded the following optimal parameters: $N = 4$, $R = 3.5$, $S = 5.0$. The values of the mutation parameters δ and m_0 were not important because the number of inner function evaluations per inner search was small (between 10 and 20) and mutation did not have time to play a part. Again the optimal value of N is close to the smallest possible value.

De Jong's F_4 function is a noisy function in 30 dimensions. A meta-optimisation to $F_4 \leq 10$ yielded the following optimal parameters: $N = 2$, $m_0 \approx 0$, $\delta = 0.00024$, $R = 0.88$, $S = 0.30$.

Shekel's (1971) function (also known as De Jong's F_5 function) is a severe multimodal test function in two dimensions, with twenty five very sharp local peaks arranged in a square 5x5 array. The global peak is located at one corner of the array. A meta-optimisation yielded the following optimal values: $N = 12$, $log_{10}(m_0)$ = -4.5, $log_{10}(\delta)$ = -4.4, $R = 1.67$, $S = 0.75$. The values of the mutation parameters δ and m_0 were at the lower end of the range, indicating that random mutation (that due to m_0) does not play a part, although the distance-dependent term $2^{-d/\delta}$ would have been important in ensuring that diversity is not lost.

In conclusion, the base algorithm of this paper proved to be robust in dealing with a variety of strandard test functions. It works best with a relatively small population size, between two and 12, because the mutation frequency is made to increase exponentially with decreasing distance between parents, to regenerate diversity after it is lost.

3.2 Identical Competing Populations

Numerous measures of efficiency have been devised by the genetic algorithm community. However, the measure most widely accepted by people interested in function optimisation is the number of fitness function evaluations, N_{eval}, required to converge to the (global) minimum, and this measure will be used in this paper.

The first step is to test whether the parallel evolution of more than one population imposes a penalty in terms of extra fitness evaluations. To this end, competitive evolutions with 1, 2, 3, 4 and 10 populations were carried out on Rosenbrock's (1960) function to $f(x) <= 0.0001$:

$$f(x) = 100 (x_2 - x_1^2)^2 + (1-x_1)^2 \tag{3}$$

This function, which is widely used in the conventional optimisation literature as well as in the evolutionary optimisation comunity, has a narrow parabolic valley with a minimum $f(1,1) = 0$. The base algorithm all used $N=12$, $m_0=0$, $\delta=0.001$, $R=1.4$, $S=0.06$ (parameter set determined from a previous optimisation by meta-evolution). For each value of N_p, the competitive evolution was carried out 200 times in order to yield statistics on the performance. Results of the single and competitive evolutions are shown in Table 1.

Table 1. Number of fitness evaluations for single ($N_p=1$) and competitive ($N_p>1$) evolutions. (Statistics from 200 runs each.)

No. of populations	1	2	3	4	10
Mean	387	394	411	408	463
Std. dev.	242	197	189	177	217
Maximum	1590	1300	1210	1210	1440
Minimum	10	110	110	120	130
Range	1580	1190	1100	1090	1310

For 2, 3 or 4 populations, there was no significant difference in the mean N_{eval}. This showed that once the "best" population is detected it monopolises the evolution and little time is wasted on the other populations. On the other hand, the spread of N_{eval} is greatly reduced, as measured by the decrease in standard deviation, range and maximum N_{eval} between $N_p=1$ and $N_p=4$. Thus, even when all parameters are kept the same, competitive evolution provides an insurance against unlucky runs.

As N_p increases to 10, however, there is a significant increase in mean N_{eval} and the spread also increase (although still showing an improvement over single evolution). It is believed that with a large number of populations, there will be several whose performance are similar and time is wasted by switching from one population to another in this fierce competition. Political scientists will no doubt draw conclusions as to the relative method of dictatorships, two-party and multiparty systems.

3.3 Competing Populations with Different Crossover Operators

At this stage, we introduce another crossover operator, the uniform crossover, which involves swapping x_i^A and x_i^B between two population members A and B for some i values. This is a variant of the classical chromosomal crossover, modified to deal with floating-point genes.

This operator can be expected to be mostly harmful in the optimisation of Rosenbrock's function, because given two members A, B lying on the floor of the parabolic valley (and therefore reasonably fit), the offsprings, which will lie in the other corners of the rectangle with diagonal AB, will be almost always on the slope and therefore less fit (Fig.1). Experimentation with 200 runs of a (single) evolution using uniform crossover 50% of the time ($p_{ux} = 0.5$) and arithmetic crossover the rest of the time shows that this is indeed the case, with mean $N_{eval}=1329$ (as compared with 387 with arithmetic crossover only).

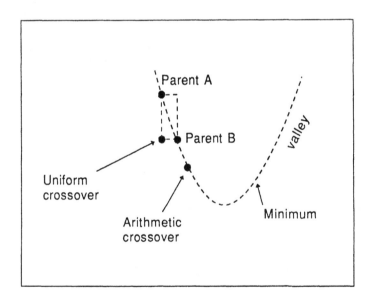

Fig. 1. Uniform vs arithmetic crossover.

How does competitive evolution deal with such a case? A competitive evolution with two populations were carried out, the first population retaining the earlier base parameters while the second uses $p_{ux}=0.5$. Over 200 runs, the mean N_{eval} obtained was 569: this is about 50% higher than the better of the single evolutions, indicating that some time did get lost due to the competition; however, it is only 43% of the worse single evolution, and much better than an arithmetic or geometric average of the two.

The same test was carried out on Shekel's (1971; quoted in de Jong, 1975) function:

$$\frac{1}{f(x)} = \sum_{j=1}^{25} \frac{1}{j + (x_1 - a_{1j})^6 + (x_2 - a_{2j})^6} \qquad (4)$$

This is a severe multimodal test function in two dimensions, with twenty five very sharp local peaks arranged in a square 5x5 array. The global peak is located at one corner of the array. The uniform crossover operator can be expected to be useful in this case, because of the regular arrangement of the peaks.

Again 200 runs each of $(N_p=1, p_{ux}=0)$, $(N_p=1, p_{ux}=0.5)$ and $(N_p=2, p_{ux}=0$ or $0.5)$ were carried out. Other parameters were: $N=2$, $m_0=0$, $\delta=0.001$, $R=0.8$, $S=0.6$. The mean N_{eval} were 1494, 1094 and 1567 respectively. In this case, because the difference between the two strategies were not great, the advantage of competitive evolution was outweighed by its disadvantage (time wasted in going from one to the other). A further cause of the performance deterioration may be that uniform crossover is more useful in the initial exploratory, peak-jumping stage of the search, while arithmetic crossover is more useful in the final polishing stage. Thus, a strategy for allowing the individual populations to swap strategies ("population crossover") may lead to further improvements.

3.4 Competing Populations with Different Sizes

Again Rosenbrock's function was used as a testbed, with population size being $N=12$ (base condition), $N=2$ (alternative) or both (for a competitive evolution). Other parameters were the same as before. The results were: single evolution with $N=12$, $N_{eval}=387$ (as previously reported); with $N=2$, $N_{eval}=1286$; and competitive evolution, $N_{eval}=709$, confirming the previous results.

3.5 Multiple Competition

To see how competitive evolution handles more than two simultaneous strategies, tests were carried out on Rosenbrock's function using the following strategies, first individually $(N_p=1)$ and then competitively $(N_p=4)$: $N=2$ or 12, $p_{ux}=0$ or 0.5. Table 2 shows that the competitive run $(N_p=4)$ is better than all but the best strategy $(N=12, p_{ux}=0)$, although there is a marked increase in N_{eval} compared with the latter.

Table 2. Number of fitness evaluations for various evolution strategies, run individually and in competition. Result from 200 runs using Rosenbrock's function.

N_p	N	p_{ux}	Av. N_{eval}
1	12	0.0	387
1	12	0.5	1329
1	2	0.0	1286
1	2	0.5	≈ 4000
4	12, 2	0.0, 0.5	1130

3.6 Effect of Population Mixing

In nature, periodic mixing of subpopulations often increases their vigour, due to the increase in the variety of the genetic pool. Such a strategy was tested on the competitive evolution method. After every cycle (involving the production of one offspring followed by a cull for each subpopulation), there is a probability $p_m=0.001$ that the fittest population and the fastest improving one (if they are separate) will randomly exchange members. Table 3 shows that in every case this leads to an improvement in the mean and the spread (as measured by the standard error in the mean). This indicates that the maximum number of fitness evaluations, due to unlucky combinations, is greatly reduced.

Table 3. Number of fitness evaluations with and without population mixing. Result from 200 runs for each strategy.

Fitness function	N_p	Competing strategies	Av. $N_{eval} \pm$ S.E.	
			No mixing	$p_m=0.001$
Rosenbrock	2	$p_{ux}=0$ or 0.5	569 ± 29	495 ± 18
Rosenbrock	2	$N=2$ or 12	709 ± 36	687 ± 30
Shekel	2	$p_{ux}=0$ or 0.5	1567 ± 82	1421 ± 74
Rosenbrock	4	$N=2$ or 12, $p_{ux}=0$ or 0.5	1130 ± 65	911 ± 39

4. Conclusions

Competitive evolution offers a variety of attractions. When using identical strategies in the subpopulations, it provides an insurance against unlucky runs while extracting only an insignificant cost in terms of extra function evaluations. When using different strategies in the subpopulations, it ensures that the best strategies are used and again the extra cost is not great. Competitive evolution is at its best when an operator - or the lack of it - has a very detrimental effect which cannot be known in advance. From an aesthetic point of view the method fits in well with the Darwinian paradigm. In fact, it is nothing more than the standard evolutionary model applied to whole populations. Occasional mixing of the best performing subpopulations may lead to further improvement.

References

Davis, L.: Adapting operator probabilities in genetic algorithms. Proc. 3rd Internat. Conf. on Genetic Algorithms, George Mason University, June 4, pp.61-69. Morgan Kaufman Pub., San Mateo, California (1989)

Davis, L., ed.: Handbook of Genetic Algorithms. Van Nostrand Reinhold, N.Y. (1991)

De Jong, K.A.: An analysis of the behavior of a class of genetic adaptive systems, Doctoral thesis, Dept. of Computer and Communication Sciences, University of Michigan, Ann Arbor (1975)

Fogel, D.B. and Atmar, J.W. (eds.): Proceedings of the First Annual Conference on Evolutionary Programming. Evolutionary Programming Society, San Diego, CA (1992)

Goldberg, D.E.: Genetic Algorithms in Search, Optimization and Machine Learning, Addison-Wesley Pub, Reading, MA (1989)

Grefenstette, J.J.: Optimization of control parameters for genetic algorithms. IEEE Trans. on Systems, Man and Cybernetics SMC-16 (1986) 122-8

Holland, J.H.: Adaptation in natural and artificial systems. MIT Press (1975).

Janikow, C.Z. and Michalewicz, Z.: An experimental comparison of binary and floating point representations in genetic algorithms. In R.K. Belew and L.B. Booker, editors, Proc. 4th Int. Conf. on Genetic Algorithms, pp.31-36. Morgan Kaufman (1991)

Michalewicz, Z.: Genetic Algorithms + Data Structures = Evolution Programs. Springer Verlag, Berlin (1992)

Rosenbrock, H.H.: An automatic method for finding the greatest or least value of a function. Computer J. 3 (1960) 175-184

Rudolph, G.: Convergence analysis of canonical genetic algorithms. IEEE Trans on Neural Networks 5 (1994) 96-101

Schaffer, J., David, R.A., Caruana, L.J. and Das, R.: A study of control parameters affecting online performance of genetic algorithms for function optimisation. In J.D. Schaffer (ed.),Proc. 3rd Internat. Conf. on Genetic Algorithms. Morgan Kaufman, San Mateo, California (1989)

Shekel, J.: Test functions for multimodal search techniques. Fifth Annual Princeton Conf. Inform. Sci. Syst. (1971)

Emergent Collective Computational Abilities in Interacting Particle Systems *

Zhong Zhang   Shuo Bai   Guojie Li

National Research Center for Intelligent Computing Systems **
Academia Sinica, P.O.Box 2704, Beijing 100080, P.R.China

Abstract. In recent years computational abilities emerging from systems in nature, especially systems studied in domains such as biology and physics, have attracted much attention from researchers in various fields. In this paper we propose a new computational model based on the interaction of charged particles, and we investigate emergent collective computational abilities of the system. In particle systems, the local motion of particles at the micro level and the state of the system as a whole on the macro level are integrated in a natural way. The local movement of particles is determined by the resultant forces acting on them, and the global system state is described by an energy function. A particle system was constructed to solve Traveling Salesman Problems (TSPs). In comparison with neural networks, this model is able to more effectively make use of the two-dimensional information of city distributions. Finally, to demonstrate the feasibility of our model, we have implemented a simulation of an interacting particle system to solve TSPs on a SUN workstation in C language. The preliminary experimental results show that there are very strong emergent collective computational abilities in interacting particle systems.

1 Introduction

In recent years, emergent collective computational behavior seen in natural systems, including those of biology, physics, evolution, and so on, has been of great interest to researchers from many disciplines. Each year there are several international conferences on this topic, and a highly interdisciplinary group of scientists from fields such as physics, biology, biochemistry, computer science, electrical engineering, economics, artificial intelligence, neuroscience, psychology, cognitive science, and so on, have begun to investigate this area from diverse viewpoints.

Perhaps the first international conference on new computational models was the 1981 Workshop on Physics and Computation held at MIT, at which Feynman first put forth the concept and proposed a model for quantum computation [12-14]. Since then there have arisen a host of novel frameworks challenging conventional ideas about the computer. In 1982, Hopfield published the first of three papers on emergent collective computational behavior in neural networks[3][4],

* This project is supported in part by National Natural Science Foundation of China.
** e-mail: zz@lumba.ncic.ac.cn

and has applied these findings in solving Traveling Salesman Problems (TSPs)[5]. This research initiated a great deal of research on neural networks, which is still ongoing. In 1975, John Holland put forth the concepts of genetic algorithms and complex adaptive systems[1], which were later continued and developed by Goldberg[2]. In 1988, Huberman developed a new computational framework involving social systems, which he calls the "ecology of computation"[15][16]. Recently, at the Santa Fe Institute, Melanie Mitchell and her colleagues have studied the evolutionary processes of emergent computation in decentralized spatially-extended systems[21-23]. All of these models, including neural networks, genetic algorithms, adaptive systems, immune systems, evolutionary processes[6][7], physics systems[8-11], the particle system presented in this paper, and others[17], are examples of nonlinear collective behavior emerging from complex natural systems. Work in this area is currently on the increase, and there is a multidisciplinary effort to find new models of computation.

This paper proposes a computational model involving a system of interacting particles, and also examines emergent collective computational behavior within the system. The local movement of a particle is determined by the resultant forces acting upon it, and the state of the system as a whole is described by an energy function, so that both local motion of particles at the micro level and the overall state of the system at macro level is organically integrated. When such a model is used to solve a specified problem, the constraints of the problem are first mapped onto the structure of the particle system. The number of particles is determined by the problem size, and the constraints are represented by the interacting forces among particles. Given an initial state, the interaction of particles can be considered an evolutionary process of collective computation. Through the interaction of particles, the system "evolves" into a stable state representing a solution to the problem. In this paper, we use such a model to attempt to solve Traveling Salesman Problems (TSPs). In comparison with neural networks, we have observed that this kind of particle system is able to better utilize the two-dimensional information of city distributions. Finally, we have carried out simulation experiments on the interacting processes in the particle system for TSPs on a SUN workstation. The results of theoretical analyses and simulation experiments show that interacting particle systems have very strong emergent collective computational abilities.

This paper is organized as follows. Section 2 presents a mechanics model for interacting particle systems, and studies the process for changing the energy of the system. Section 3 establishes an interacting particle system for solving TSPs, and describes its implementation on a SUN workstation in C language. Section 4 analyzes some preliminary experimental phenomena and discusses the results. Section 5 outlines the research significance of this paper, and leaves several open problems to be solved in the near future. Finally, a conclusion is given in Section 6.

2 The Basic Model for Interacting Particle Systems

Suppose that in an interacting particle system consisting of N particles, there is an interaction force between each two particles, particle i and particle j, $\mathbf{F}_{ij}$. The resultant forces on particle i is:

$$\mathbf{F}_i = \sum_{\substack{j \neq i}}^{N} \mathbf{F}_{ij} \qquad (i, j = 1, 2, \cdots, N) \qquad (1)$$

Based on Newton's Second Law, particle i will move under the action of resultant force $\mathbf{F}_i$, and its dynamic equation is:

$$\mathbf{F}_i = m_i \mathbf{a}_i = m_i \frac{d\mathbf{v}_i}{dt} \qquad (i = 1, 2, \cdots, N) \qquad (2)$$

here, m_i is the mass of particle i, $\mathbf{a}_i$ is the acceleration of particle i, and $\mathbf{v}_i$ is the velocity of particle i. $\frac{d\mathbf{v}_i}{dt}$ is the derivative of velocity with respect to time. Particle i has potential energy:

$$E_i^p = -\mathbf{F}_i \cdot \mathbf{r}_i = -W_i \qquad (i = 1, 2, \cdots, N) \qquad (3)$$

in which, $\mathbf{r}_i$ is the position vector of particle i. W_i is the amount of work used in moving particle i to position $\mathbf{r}_i$. We define the potential energy as the maximum in infinite distance.

In the following, we will discuss the process of changing the potential energy of particle i when its displacement is changed with respect to time t.

$$\frac{dE_i^p}{dt} = -\mathbf{F}_i \cdot \frac{d\mathbf{r}_i}{dt} = -\mathbf{F}_i \cdot \mathbf{v}_i \qquad (i = 1, 2, \cdots, N) \qquad (4)$$

where, $\frac{d}{dt}$ is the derivative with respect to time.

Substitute Eq.(2) into Eq.(4), we get

$$\frac{dE_i^p}{dt} = -m_i \frac{d\mathbf{v}_i}{dt} \cdot \mathbf{v}_i = -\frac{1}{2} m_i \left(\frac{dv_i^2}{dt} \right) \qquad (i = 1, 2, \cdots, N) \qquad (5)$$

in which, v_i is the absolute value of velocity vector $\mathbf{v}_i$.

The potential energy of the entire system is changed with respect to time t as follows:

$$\frac{dE^p}{dt} = \frac{d\sum_i^N E_i^p}{dt} = -\frac{1}{2} \sum_i^N m_i \frac{d}{dt} v_i^2 \qquad (6)$$

In fact, this is a Hamiltonian conservation system. The potential energy is changed into the kinetic energy, and the total energy of the system is invariant. i.e.,

$$\frac{dE}{dt} = \frac{d(E^p + E^v)}{dt} = \sum_i^N \frac{d}{dt} \left(E_i^p + \frac{1}{2} m_i v_i^2 \right) = 0 \qquad (7)$$

In the beginning, the kinetic energy of the system is very small or 0 ($v_i \rightarrow 0$). Under the action of forces generated by the potential energy, the kinetic energy is increased step-by-step. When particle i moves to a special place, we fix it at that point and release its kinetic energy, at which point the potential energy of the system is decreased. When all particles in the system are not moving, $v_i = 0$ ($i = 1, 2, ..., N$). The system arrives at a stable state, and has a lower potential energy. This stable state corresponds to a solution of the problem to be solved.

The interaction among particles under the resultant forces "represents" a kind of emergent collective computation within such particle systems.

3 Interacting Particle System for TSPs

3.1 Model of interacting particle system for TSPs

Let us construct an interacting particle system to solve TSPs. For a N-city TSP, N negatively-charged particles are located at corresponding coordinate positions of N cities, and N moving particles with positive charges are linked in a closed circle by an elastic line. In an interacting particle system of $2N$ particles, there are in all three kinds of interacting forces.

The first kind of force is attraction $\mathbf{F}_{ij}^a$ between moving particle i with positive charge and fixed negatively-charged particle j. According to Coulomb's Law concerning static electricity, this is

$$\mathbf{F}_{ij}^a = -K_a \frac{Q^2}{|\mathbf{r}_i - \mathbf{c}_j|^2} \cdot \hat{e}_{ij} \qquad (i, j = 1, 2, \cdots, N) \qquad (8)$$

in which, $\mathbf{r}_i$ is the position vector of moving particle i, and $\mathbf{c}_j$ is the position vector of city j. $\hat{e}_{ij}$ is a unit vector in the direction of particle i to city j. Q is electric charge, and K_a is the Coulomb constant. $|\mathbf{x}|$ indicates the absolute value of vector $\mathbf{x}$.

The second kind of force is mutual repulsion $\mathbf{F}_{lk}^r$ between moving positively-charged particle l and moving positively-charged particle k, which is

$$\mathbf{F}_{lk}^r = K_r \frac{Q^2}{|\mathbf{r}_l - \mathbf{r}_k|^2} \cdot \hat{r}_{lk} \qquad (l, k = 1, 2, \cdots, N) \qquad (9)$$

in which, $\mathbf{r}_l$ is the position vector of moving particle l, $\mathbf{r}_k$ is the position vector of moving particle k. $\hat{r}_{lk}$ is a unit vector in the direction of particle l to particle k. Q is electric charge, and K_r is the Coulomb constant.

The third kind of force is elasticity $\mathbf{F}_{m,m\pm1}^e$ between moving particle m and the neighboring moving particle $(m \pm 1)$, which is

$$\mathbf{F}_{m,m\pm1}^e = K_e[|\mathbf{r}_m - \mathbf{r}_{m\pm1}| - \Delta] \cdot sgn[|\mathbf{r}_m - \mathbf{r}_{m\pm1}| - \Delta] \cdot \hat{r}_{m,m\pm1} \qquad (10)$$

in which, $\mathbf{r}_m$ is the position vector of moving particle m, $\mathbf{r}_{m\pm1}$ is the position vector of the neighboring moving particle, $\hat{r}_{m,m\pm1}$ is the unit vector, and Δ is a given length of elastic line. K_e is the elastic factor satisfied by Hooke's Law of elasticity. Function $sgn(x)$ is defined as follows.

$$sgn(x) = \begin{cases} 1 & x > 0 \\ 0 & x \leq 0 \end{cases}$$

N negative charges produce a gravitational field caused by N cities in the TSP. N positively-charged particles are moving in the gravitational field, and are constrained by the elastic attraction forces of neighboring particles upon each other, which indicates constraints to be satisfied in the TSP. There is a requirement that the length of a closed path of a TSP be as small as possible. When a moving positively-charged particle i coincides with the negative charge located at the coordinate position of city j, this pair of charges cancel each other out and disappear, and the moving particle i stops at that point. Under action of the three kinds of forces, after all moving particles with positive charges completely counteract the corresponding negative charges and stop, the entire system arrives at a stable state which represents a solution to the TSP to be solved. In this stable state, N positively charged particles constitute a closed cycle corresponding to the positions of N cities, giving an efficient route to visit N cities.

In the system, once particles have interacted sufficiently through the three kinds of forces among them, the system finally arrives at a state of equilibrium in which all particles stop.

In the following section, we will examine collective computational abilities emerging from interacting particle systems applied to several TSPs.

3.2 Simulation of an interacting particle system for TSPs

We have implemented a simulation system of the interacting particle system to solve TSPs on a SUN workstation in C language. The programming design framework of this particle system is shown in Fig.1. First, the 10-city TSP[5] is used as an example. In the simulation experiments, the parameters chosen for the particle system were: $m_i = 1, Q = 1, K_a = 1, K_r = 1, K_e = 2$, difference of iterative time being $\Delta t = 0.001$. An initial state of the particle system is given as follows.

First, we determine the coordinate of the mass center of city distributions. Then, the average radius $\overline{R}$ is taken as the average value of the distance from a city position to the coordinate of the mass center. N particles are distributed equally on the circumference, whose center is the mass center and whose radius is the average radius $\overline{R}$. These particles are also linked into a N-edge polygon by an elastic line, in which Δ is about $\frac{2\pi}{N}\overline{R}$. Given an initial state of the interacting particle system, shown in Fig.2(a) (where, $N = 10$), under the action of three kinds of forces, N particles with positive charge begin to move in the system. Resultant forces on a particle generate an acceleration, which leads to change in the velocity of a moving particle, and causes a displacement. In the movement process, if the distance between two neighboring particles linked with elasticity line is more than Δ, then there is an elastic force between them, but if the distance between the two neighboring particles is not more than Δ, then there is not. When a moving particle coincides with some city coordinate, then it will stop at that point, shown in Fig.2(b). Particles move consecutively this way under the interactions among them (shown in Fig.2(c)), until all positive charges cancel out with an equal number of negative charges, at which point all moving particles stop. At this moment, the system has already arrived at a stable state, shown in Fig.2(d). This corresponds to a solution of the 10-city TSP. The path length L is 2.78, which is a quite good solution inferior only to the optimal solution ($L = 2.71$).

In this section, we use only a small-scale example, the 10-city TSP, to test our model and method, and to demonstrate its feasibility. We will give preliminary experimental results of the 30-city and 50-city problems in the next section.

4 Experimental Results and Discussions

From the above simulation experiments, it is not difficult to arrive at quite satisfactory solutions using an interacting particle system to solve TSPs. This system is able to sufficiently utilize the two-dimensional information of city position distributions, which can produce a two-dimensional effects dynamically in the motion of interacting particles. It should be mentioned that in simulation experiments, the system parameters, such as K_a, K_r, K_e and so on, are easily tuned. It is clear that this interacting particle system is concise, highly efficient, and easily implemented.

In simulation experiments, we have observed several phenomena.

1. When interaction parameters do not match, there may occur oscillations in the movement process, preventing the particle system from converging to a stable state. However, these kinds of oscillations may be easily eliminated if we change the parameters using a simple technique.

2. In interacting particle systems, the motion states of particles are described by position vector r_i, velocity of particle v_i, etc. When the displacement of a particle is changed, the total energy of the system is changed, such that two-dimensional effects occur in the interacting particle system. This is rather different from the previous Hopfield neural networks, in which the state variable of the system is the neuron state V_i, a scalar quantity. Therefore, there are no two-dimensional effects in the dynamic procedure of Hopfield neural networks, and the two-dimensional information of the distribution of city positions cannot be utilized to guide the evolutionary process of the system, but may merely be helpful in selecting an initial state of the neural networks.

3. The elastic approach is a kind of pattern matching technique[18], which was first used in character recognition. In a letter to *Nature*[19], Durbin and Willshaw applied this approach to solve TSPs and obtained quite satisfactory solutions. The elastic net approach is also able to utilize the two-dimensional information of geometrical structures[20]. There are some similarities between the elastic nets and the interacting particle system proposed in this paper in solving TSPs. Both are able utilize the two-dimensional structure information of city distributions. However, their practical problem-solving strategies are different. We will explore combining these two approaches to solving TSPs in the future.

4. The interacting particle system is a Hamiltonian system, in which kinetic energy and potential energy are transformed each other, and the total energy is conserved. In the motion of particles, the total energy of the system does not

change. However, when a positively-charged particle collides with a negatively-charged particle, i.e., the moving particle coincides with the position of a city coordinate, then the two charges cancel each other out at that point. The particle is thus not movable, and its kinetic energy is released suddenly. In theory, this part of the kinetic energy is transformed into thermal energy or other forms of energy, but we consider only mechanical energy here. Therefore, the potential energy of the system is decreased relatively, such that the system terminates at a lower-energy stable state.

The Hamiltonian system guarantees the convergence and the stability of interacting particle systems, and Hopfield's neural networks are energy dispersive systems, in which the energy continually decreases in the dynamic process.

5. In interacting particle systems, discrete combinatorial optimization problems are transformed into a continuous constrained system in a natural way. This is somewhat different from Hopfield neural networks, in which the system is forced back into the discrete state space from continuous state space by the use of a nonlinear threshold function to obtain a solution to the problem.

6. In fact, a position vector may be decomposed into the superposition of two scalar quantities, $\mathbf{r} = x\mathbf{i} + y\mathbf{j}$. The motion of a particle is the result of its motion along x axis and y axis. In practice, this is to some degree equivalent to a kind of coupling effect that exists between two Hopfield networks, one x network and another y network. Of course, this coupling is implicit within the system. Therefore, in the future, the interacting particle system may be hardware implemented by use of a coupling network of two circuit networks.

We use certain techniques to improve the performance of the particle system. First, we consider damping effects of a particle in motion. The particle motion equation is:

$$\frac{d\mathbf{v}_i}{dt} = \frac{\mathbf{F}_i}{m_i} - \alpha\mathbf{v}_i \qquad (i = 1, 2, \cdots, N) \tag{11}$$

in which, α is a damping factor. Eq.(7) is modified into

$$\frac{dE}{dt} = -\sum_i^N \frac{1}{2}\alpha m_i v_i^2 \tag{12}$$

here, $\alpha > 0$, $m_i > 0$ and $v_i^2 > 0$. Therefore, when $\frac{dE}{dt} \leq 0$, the energy of the system is released slowly. $\frac{dE}{dt} = 0$ implies that $v_i = 0$ for all i. So the motion seeks the minima of E and stops at such points. It seems as if the temperature parameter T is decreased slowly in simulated annealing.

Secondly, when the system terminates at a stable state, if it is not a global optimal point, then we may adopt the perturbation approach stochastically to introduce tiny deviations in the velocity and the position of particles. The system then starts to move under the interaction among particles, and at the same time we dynamically tune system parameters. For example, the elastic force between two neighboring particles is enhanced gradually.

After these techniques are used, we obtain two optimal paths for the 10-city TSP in experiments, shown in Fig.3(a) and Fig.3(b). Through many perturbations of the particle system, we get two optimal paths for the 30-city TSP, shown in Fig.4(a) and Fig.4(b). Fig.5 gives the evolutionary process of the interacting particle system for TSPs in simulation experiments, in which L is the path length of TSP and *Times* is the number of random perturbations of the particle system.

5   Research Significance and Open Problems

Computational abilities emerging from natural systems is a new exciting research direction. The interacting particle system proposed in this paper is a new kind of computational model for problem solving inspired by systems in nature. In the system, the interaction among particles is a kind of cooperative effect, one particle being considered as one agent. We have studied the motion process of mutually interacting particles, as well as the evolutionary process of the entire system. Because the state variable of particles is position vector r, our approach is a kind of two-dimensional computational model. The particle system is an open system, and many heuristics and advantages of other methods may easily be added to the system to improve its performance and to enhance its problem-solving efficiency.

There are some open problems with regards to our method, and the following are several research directions.

1. In this paper, we use only small-scale examples, such as the 10-city and 30-city problems, to test our model and method, and to demonstrate its feasibility. We intend later to use this model to solve larger-scale TSPs[7], involving 50, 75, 100, and 1000 cities, in order to observe the performance of the system.

2. In theory, we will analyze quantitatively the optimal ratio among system parameters K_a, K_r, K_e, and their affects on the performance of particle systems, as well as their influence on their dynamic behavior.

3. The model and method proposed in this paper may be more suitable for TSPs whose city distributions are convex (such as the 10-city and 30-city problems). If the city distribution is concave (such as the 50-city and 75-city problems[7]), then we may first divide the problem into several segments such that the local city distribution of each sub-problem is convex, and then use the model to solve these sub-problems. Finally, we combine solutions of sub-problems into a solution for the TSP.

4. We will explore the possibility of combining the interacting particle system with other advanced techniques, for example, using genetic algorithms to select a better initial state of the system, as well as combining the system with the elastic net approach in order to improve its performance.

6   Conclusion

In this paper, we attempt to study the collective computational abilities emerging from particle systems.

We propose a new computational model, the interacting particle system, and investigate emergent collective computational abilities in the system. We construct a kind of particle system to solve TSPs. In comparison with neural networks, we have observed that this model is able to more effectively utilize the two-dimensional information of city distributions. Finally, we have implemented a simulation of interacting particle systems to solve TSPs on a SUN workstation in C language. The preliminary experimental results show that there are very strong emergent collective computational abilities in interacting particle systems.

Some open problems are given in the end of the above section, the related research results will be reported in another paper in the near future.

References

1. Holland, J.: ADAPTATION IN NATURAL AND ARTIFICIAL SYSTEMS. Ann Arbor, The University of Michigan Press (1975)
2. Goldberg, D.: GENETIC ALGORITHMS IN SEARCH, OPTIMIZATION, AND MACHINE LEARNING. Addison-Wesley Publishing Company (1989)
3. Hopfield, J.: Neural networks and physical systems with emergent collective computational abilities. Proc. Natl. Acad. Sci. USA. **79** (1982) 2554–2558
4. Hopfield, J.: Neurons with graded response have collective computational properties like those of two-state neurons. Proc. Natl. Acad. Sci. USA. **81** (1984) 3088–3092
5. Hopfield, J.: Neural computation of decisions in optimization problems. Biological Cybernetics. **52** (1985) 141–152
6. Fogel, D.: An evolutionary approach to the Traveling Salesman Problem. Biol. Cybernet. **60**(2) (1988) 139–144
7. Fogel, D.: Applying evolutionary programming to selected Traveling Salesman Problems. Cybernetics and Systems: An International Journal. **24** (1993) 27–36
8. Wheeler, J.: Information, physics, quantum: the search for links. Proc. of the 1988 Workshop on Complexity, Entropy, and the Physics of information, Santa Fe, USA. Addison-Wesley Publishing Company (1990) 3–28
9. Davies, P.: Why is the physical world so comprehensible. Proc. of the 1988 Workshop on Complexity, Entropy, and the Physics of information, Santa Fe, USA. Addison-Wesley Publishing Company (1990) 61–70
10. Toffoli, T.: What are nature's 'Natural' ways of computing ? Proc. of the Workshop on Physics and Computation, Dallas, USA. IEEE Computer Society Press (1993) 5–9
11. Leão, L.: Artificial Physics: the soul of a new discipline. Proc. of the Workshop on Physics and Computation, Dallas, USA. IEEE Computer Society Press (1993) 10–20
12. Feynman, R.: Simulating physics with computers. International Journal of Theoretical Physics. **21** (1982) 467–488
13. Feynman, R.: Quantum mechanical computers. Foundations of Physics. **16** (1986) 507–531
14. Feynman, R.: Tiny computers obeying quantum mechanical laws. New Directions in Physics: The Los Alamos 40th Anniversary **7** (1987) 7–25
15. Huberman, B.: The ecology of computation. The Ecology of Computation. Elsevier Science Publishers B.V. (1988) 1–4
16. Huberman, B., Hogg, T.: The behavior of computational ecologies. The Ecology of Computation. Elsevier Science Publishers B.V. (1988) 77–115

17. Farmer, J.: EVOLUTION, GAMES, AND LEARNING: Models for Adaptation in Machines and Nature. Physica **22D** (1986) vii–xii
18. Widrow, B.: The rubber mask technique, Parts I and II. Pattern Recognition **5** (1973) 175–211
19. Durbin, R., Willshaw, D.: An analogue approach to the travelling salesman problem using an elastic net method. Nature. **326** (1987) 689–691
20. Durbin, R., Szeliski, R., Yuille, A.: An analysis of the elastic net approach to the travelling salesman problem. Neural Computation. **1** (1989) 348–358
21. Mitchell, M., Crutchfield, J., Hraber, P.: Evolving cellular automata to perform computations: mechanisms and impediments. Physica D (1994) (to appear)
22. Das, R., Mitchell, M., Crutchfield, J.: A genetic algorithm discovers particle-based computation in cellular automata. Third Parallel Problem-Solving From Nature Conference. (1994) (to appear)
23. Crutchfield, J., Mitchell, M.: The evolution of emergent computation. Science. (1994) (to appear)

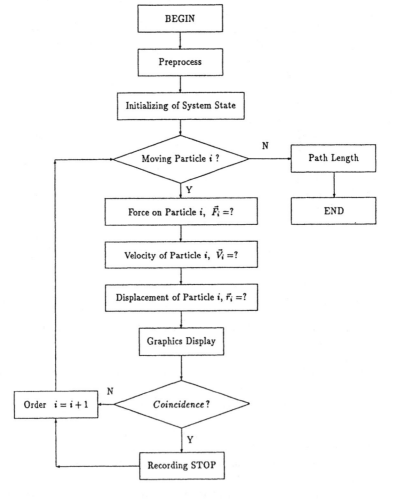

Figure 1: Simulation Implementation of Interacting Particle System

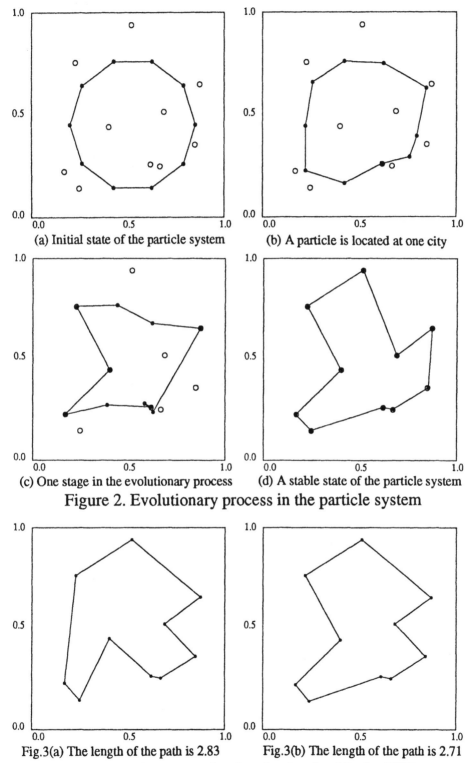

(a) Initial state of the particle system

(b) A particle is located at one city

(c) One stage in the evolutionary process

(d) A stable state of the particle system

Figure 2. Evolutionary process in the particle system

Fig.3(a) The length of the path is 2.83

Fig.3(b) The length of the path is 2.71

Figure 3. The two optimal paths for the 10-city TSP

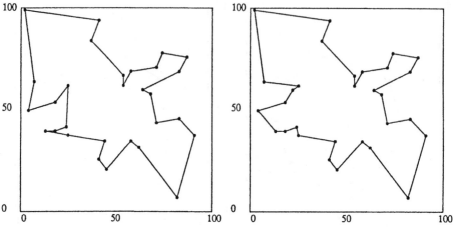

Fig.4(a) The length of the path is 434.375   Fig.4(b) The length of the path is 427.855

Figure 4. The two optimal paths for the 30-city TSP

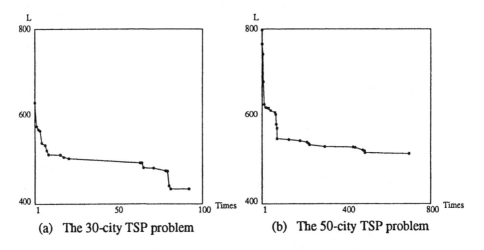

(a) The 30-city TSP problem   (b) The 50-city TSP problem

Figure 5. Evolutionary optimization of the best tour in the TSP problems

A Perspective on Evolutionary Computation

Zbigniew Michalewicz

Department of Computer Science
University of North Carolina
Charlotte, NC 28223, USA

Abstract. During the last three decades there has been a growing interest in algorithms which rely on analogies to natural processes. The emergence of massively parallel computers made these algorithms of practical interest. The best known algorithms in this class include evolutionary programming, genetic algorithms, evolution strategies, simulated annealing, classifier systems, and neural networks.

In this paper we discuss a subclass of these algorithms—those which are based on the principle of evolution (survival of the fittest). A common term, recently accepted, refers to such techniques as 'evolutionary computation' methods.

The paper presents a perspective of the field of evolutionary computation. It discusses briefly the concept of evolutionary computation, presents the author's first experience with these methods, provides a discussion on relationship between evolutionary computation techniques and the problem specific knowledge, and identifies some current critical issues.

1 Introduction

The field of evolutionary computation is approaching a stage of maturity. There are several, well established international conferences that attract hundreds of participants (International Conferences on Genetic Algorithms—ICGA [27, 28, 44, 6, 23], Parallel Problem Solving from Nature—PPSN [48, 32], Annual Conferences on Evolutionary Programming—EP [19, 20, 49]); new annual conferences are getting started, e.g., IEEE International Conferences on Evolutionary Computation [41]. Also, there are many workshops, special sessions, and local conferences every year, all around the world. A new journal, *Evolutionary Computation* (MIT Press) [7], is devoted entirely to evolutionary computation techniques; many other journals organized special issues on evolutionary computation (e.g., [17, 35]). Many excellent tutorial papers [4, 5, 42, 52, 18] and technical reports provide more-or-less complete bibliographies of the field [1, 26, 43, 38]. There is also *The Hitch-Hiker's Guide to Evolutionary Computation* prepared initially by Jörg Heitkötter and currently by David Beasley [29], available on comp.ai.genetic interest group (Internet), and a new text, *Handbook of Evolutionary Computation*, is currently being prepared [3].

This paper discusses two issues related to evolutionary computation. First, a connection between the problem specific knowledge, representation of individuals, operators and the evolutionary computation techniques is explored to some degree; this part provides ground for further discussion on similarities and differences between major evolutionary techniques, like genetic algorithms (GAs), evolutionary strategies (ES), and evolutionary programming (EP). Second, we concentrate on the current trends in the field.

The paper is organized as follows. The next section discusses briefly the concept of evolutionary computation. Section 3 presents the author's first experience with evolutionary techniques, and Section 4 provides a discussion on relationship between evolutionary computation techniques and the problem specific knowledge. The final section identifies some current critical issues in the field.

2   Evolutionary computation

Evolutionary computation algorithms are based on the principle of evolution (survival of the fittest). In these algorithms a population of individuals (potential solutions) undergoes a sequence of unary (mutation type) and higher order (crossover type) transformations. These individuals strive for survival: a selection scheme, biased towards fitter individuals, selects the next generation. After some number of generations, the program converges—the best individual represents a near-optimum solution.

The two most important issues in the evolution process are population diversity and selective pressure. These factors are strongly related: an increase in the selective pressure decreases the diversity of the population, and vice versa. In other words, strong selective pressure "supports" the premature convergence of the search and a weak selective pressure can make the search ineffective. Different evolutionary techniques use different scaling methods and different selection schemes (e.g., proportional selection, ranking, tournament) to strike a balance between these two factors.

However, the structure of any evolutionary computation algorithm is very much the same; a sample structure is shown in Figure 1.

```
procedure evolutionary algorithm
begin
    t ← 0
    initialize  P(t)
    evaluate  P(t)
    while (not termination-condition) do
    begin
        t ← t + 1
        select  P(t) from  P(t − 1)
        alter  P(t)
        evaluate  P(t)
    end
end
```

Fig. 1. The structure of an evolutionary algorithm

The evolutionary algorithm maintains a population of individuals, $P(t) = \{x_1^t, \ldots, x_n^t\}$ for iteration t. Each individual represents a potential solution to the problem at hand, and is implemented as some data structure S. Each solution x_i^t is evaluated to give some measure of its "fitness". Then, a new population (iteration $t+1$) is formed by selecting the more fit individuals (select step). Some members of the new population undergo transformations (alter step) by means of "genetic" operators to form new solutions. There are unary transformations m_i (mutation type), which create new individuals by a small change in a single individual ($m_i : S \rightarrow S$), and higher order transformations c_j (crossover type), which create new individuals by combining parts from several (two or more) individuals ($c_j : S \times \ldots \times S \rightarrow S$). After some number of generations the algorithm converges—it is hoped that the best individual represents a near-optimum (reasonable) solution.

The data structure S used for a particular problem and a set of 'genetic' operators constitute the most essential components of any evolutionary algorithm. For example, the original genetic algorithms, devised to model *adaptation processes*, mainly operated on binary strings and used recombination operator with mutation as a background operator [30]. Evolution strategies [47] used real variables[1] and aimed at *numerical optimization*. Because of that, the individuals incorporated also a set of strategic parameters. Evolution strategies relied mainly on mutation operator (Gaussian noise with zero mean); only recently a discrete recombination was introduced for object variables and intermediate recombination—for strategy parameters. On the other hand, evolutionary programming techniques [22] aimed at building a system to solve *prediction tasks*. Thus they used a finite state machine as a chromosome and 5 mutation operators (change the output symbol, change a state transition, add a state, delete a state, or change the initial state).[2]

In the next section we discuss a particular evolutionary system which is hard to classify into categories of genetic algorithms, evolution strategies or evolutionary programming. The system was developed for the transportation problem and it uses matrix representation for chromosomes with a problem-specific mutation and arithmetical crossover. It is unclear whether the system can be classified as genetic algorithm without any support of scheme theorem or building-block hypothesis. It uses floating point representation (as evolution strategies do), however, it does not incorporate any control parameters into the structure of its chromosomes. The matrix representation may suggest the case of evolution programming since finite state machines were represented as matrices, however, the problem is clearly 'to optimize' and not 'to predict'. After discussing the features of the evolutionary system for the transportation problem (next section) we will return to the general discussion on similarities and differences between various techniques from the perspective of incorporating problem-specific knowledge into these algorithms (Section 4).

---

[1] However, they started with integer variables as an experimental optimum-seeking method.

[2] 'Change', 'add', and 'delete' are specialized versions of mutation.

3 Evolutionary algorithm for the transportation problem

In 1991 two papers were published on GA-based systems for the transportation problem: for linear [51] and nonlinear [36] cases. To the best of the author's knowledge they were the first GA-based system to use non-string chromosome structures;[3] specialized 'genetic' operators were introduced to preserve feasibility of solutions. This section summarizes this research by stating the problem and discussing possible representations and operators.

The transportation problem is one of the simplest constrained optimization problems that have been studied. It seeks the determination of a minimum cost transportation plan for a single commodity from a number of sources to a number of destinations. A destination can receive its demand from one or more sources. The objective of the problem is to determine the amount to be shipped from each source to each destination such that the total transportation cost is minimized.

If the transportation cost on a given route is directly proportional to the number of units transported, we have a *linear transportation problem*. Otherwise, we have a *nonlinear transportation problem*.

Assume there are n sources and k destinations. The amount of supply at source i is $source(i)$ and the demand at destination j is $dest(j)$. The cost of transporting flow x_{ij} from source i to destination j is given as a function f_{ij}. Thus the total cost is a separable function of the individual flows rather than interactions between them. The transportation problem is given as:

$$\text{minimize } total = \sum_{i=1}^{n} \sum_{j=1}^{k} f_{ij}(x_{ij})$$

subject to

$$\sum_{j=1}^{k} x_{ij} \leq source(i), \text{ for } i = 1, 2, \ldots, n,$$
$$\sum_{i=1}^{n} x_{ij} \geq dest(j), \text{ for } j = 1, 2, \ldots, k,$$
$$x_{ij} \geq 0, \text{ for } i = 1, 2, \ldots, n \text{ and } j = 1, 2, \ldots, k.$$

The first set of constraints stipulates that the sum of the shipments from a source cannot exceed its supply; the second set requires that the sum of the shipments to a destination must satisfy its demand.

The above problem implies that the total supply $\sum_{i=1}^{k} source(i)$ must at least equal total demand $\sum_{j=1}^{n} dest(j)$. When total supply equals total demand (total flow), the resulting formulation is called a *balanced* transportation problem. It differs from the above only in that all constraints are equations; that is,

$$\sum_{j=1}^{k} x_{ij} = source(i), \text{ for } i = 1, 2, \ldots, n,$$
$$\sum_{i=1}^{n} x_{ij} = dest(j), \text{ for } j = 1, 2, \ldots, k.$$

In constructing an evolutionary algorithm for the transportation problem, a selection of appropriate data structure S together with the set of appropriate 'genetic' operators is of utmost importance. It is because there are several hard constraints to be satisfied. As stated by Davis [11]:

---

[3] Fogel, Owens & Walsh [21] in their EP technique represented finite state machines as matrices.

"Constraints that cannot be violated can be implemented by imposing great penalties on individuals that violate them, by imposing moderate penalties, or by creating decoders of the representation that avoid creating individuals violating the constraint. Each of these solutions has its advantages and disadvantages. If one incorporates a high penalty into the evaluation routine and the domain is one in which production of an individual violating the constraint is likely, one runs the risk of creating a genetic algorithm that spends most of its time evaluating illegal individuals. Further, it can happen that when a legal individual is found, it drives the others out and the population converges on it without finding better individuals, since the likely paths to other legal individuals require the production of illegal individuals as intermediate structures, and the penalties for violating the constraint make it unlikely that such intermediate structures will reproduce. If one imposes moderate penalties, the system may evolve individuals that violate the constraint but are rated better than those that do not because the rest of the evaluation function can be satisfied better by accepting the moderate constraint penalty than by avoiding it. If one builds a "decoder" into the evaluation procedure that intelligently avoids building an illegal individual from the chromosome, the result is frequently computation-intensive to run. Further, not all constraints can be easily implemented in this way."

There are other possibilities as well. Sometimes it is worthwhile to design a repair algorithm which would 'correct' an infeasible solution into a feasible one. In such cases, there is an additional question to be resolved: it is whether the repaired chromosome should replace the original one in the population, or rather the repair process is run only for evaluation purpose.[4] Also, there is a possibility of using a data structure appropriate for the problem at hand together with the set of specialized operators; such approach was described in detail in [34]. We will examine briefly these possibilities in turn.

It is possible to build a "classical" genetic algorithm for the transportation problem, where chromosomes (i.e. representation of solutions) are bit strings—lists of 0's and 1's. A straightforward approach is to create a vector $(v_1, v_2, \ldots, v_p)$ $(p = n \cdot k)$, such that each component v_i $(i = 1, 2, \ldots, p)$ is a bit vector $\langle w_0^i, \ldots, w_s^i \rangle$ representing a value associated with row j and column m in the allocation matrix, where $j = \lfloor (i-1)/k + 1 \rfloor$ and $m = (i-1) \bmod k + 1$.

However, it is difficult to design a meaningful set of penalty functions. If penalty functions are moderate, the system often returns [33] an infeasible solution $x_{ij} = 0.0$ for all $1 \leq i \leq n$, $1 \leq j \leq k$, which yields the "optimum" transportation cost (zero)! It seems that high penalties have much better chances to

---

[4] Recently Orvosh and Davis [39] reported so-call 5% rule which states that if replacing original chromosomes with a 5% probability, the performance of the algorithm is better than if replacing with any other rate. In particular, it is better than with 'never replacing' or 'always replacing' strategies. However, the rule has some exceptions [34].

force solutions into a feasible region of the search space, or at least to return solutions which are 'almost' feasible. However, it should be stressed that

- with high penalties very often the system would settle for the first feasible solution found, or
- if a solution is 'almost' feasible, the process of finding a 'good' correction can be quite complex for high dimensional problems. We can think about this step as a process of solving a new transportation problem with modified marginal sums (which represent differences between actual and required totals), where variables, say, δ_{ij}, represent respective corrections to original variables x_{ij}.

We conclude that the penalty function approach is not the most suitable for solving constrained problems of this type.

Judging from the previous paragraph it should be clear that the 'repair algorithm' approach has also slim chances to succeed. Even if the initial population consists of feasible solutions only, there are some serious difficulties. For example, let us consider a required action when a feasible solution undergoes mutation. The mutation is usually defined as a change in a single bit in a solution-vector. This would correspond to a change of one value, v_i. This, in turn, would trigger a series of changes in different places (at least 3 other changes) in order to maintain the constraint equalities (note also, that we always have to remember in which column and row a change was made—despite a vector representation we think and operate in terms of rows and columns).

There are some other open questions as well. Assume that two random points (v_i and v_m, where $i < m$) are selected such that they do not belong to the same row or column. Let us assume that v_i, v_j, v_k, v_m ($i < j < k < m$) are components of a solution-vector (selected for mutation) such that v_i and v_k as well as v_j and v_m belong to a single column, and v_i and v_j as well as v_k and v_m belong to a single row. That is, in matrix representation:

$$
\begin{array}{ccccc}
\cdots & \cdot & \cdots & \cdot & \cdots \\
\cdots & \cdot & \cdots & \cdot & \cdots \\
\cdots & v_i & \cdots & v_j & \cdots \\
\cdots & \cdot & \cdots & \cdot & \cdots \\
\cdots & \cdot & \cdots & \cdot & \cdots \\
\cdots & v_k & \cdots & v_m & \cdots \\
\cdots & \cdot & \cdots & \cdot & \cdots \\
\cdots & \cdot & \cdots & \cdot & \cdots
\end{array}
$$

Now in trying to determine the smallest change in the solution vector we have a difficulty. If we increase the value v_i by a constant C we have to decrease each of the values v_j and v_k by the same amount. What happens if $v_j < C$ or $v_k < C$? We could set $C = min(v_i, v_j, v_k)$, but then most mutations would result in no change, since the probability of selecting three non-zero elements would be close to zero for solutions from the surface of the simplex. Thus methods involving single bit changes result in inefficient mutation operators with complex expressions for checking the corresponding row or column of the selected element.

The situation is even worse if we try to define a crossover operator. Breaking a vector at a random point can result in the appearance of numbers v_i larger than all $sour(i)$ and $dest(j)$, obviously violating constraints. Even if we design a method to provide that all numbers in the solution-vectors of offspring resulting from crossovers are in a reasonable range it is more than likely that these new solutions would still violate the constraints. If we try to modify these solutions to obey all constraints we would then lose all similarities with the parents. Moreover, the way to do this is far from obvious. We conclude that the repair algorithm approach is not the most suitable for solving constrained problems of this type.

The third possibility, the use of decoders, is almost out of question. Decoders are used mainly for discrete optimization problems (e.g., knapsack problem, see [34]), and it might be difficult to design a decoder scheme for continuous case. A GA-based system based on decoders was built for the linear case of the transportation problem [51], but is impossible to use the proposed decoder for the nonlinear case [36]. It was clear that decoders were not the most suitable for solving constrained problems of this type.

The general conclusion from the above discussion is that the vector representation (whether used with penalty functions, repair algorithms, or decoders) is not the best data structure for the transportation problem. Perhaps the most natural representation of a solution for this problem is a two dimensional structure. After all, this is how the problem is presented and solved by hand. In other words, a matrix $V = (x_{ij})$ $(1 \leq i \leq k, \ 1 \leq j \leq n)$ may represent a solution; each x_{ij} is a real number.

It is relatively easy to initialize a population so it contains only feasible individuals. In [36] a particular *initialization* procedure is discussed which introduces as many zero elements as possible. Such *initialization* procedure can be used to define a set of 'genetic' operators (two mutations and one crossover) which would preserve feasibility of solutions:

mutation-1. Assume that $\{i_1, i_2, \ldots, i_p\}$ is a subset of $\{1, 2, \ldots, k\}$, and $\{j_1, j_2, \ldots, j_q\}$ is a subset of $\{1, 2, \ldots, n\}$ such that $2 \leq p \leq k$, $2 \leq q \leq n$.
Denote a parent for mutation by the $(k \times n)$ matrix $V = (x_{ij})$. Then we can create a $(p \times q)$ submatrix $W = (w_{ij})$ from all elements of the matrix V in the following way: an element $x_{ij} \in V$ is in W if and only if $i \in \{i_1, i_2, \ldots, i_p\}$ and $j \in \{j_1, j_2, \ldots, j_q\}$ (if $i = i_r$ and $j = j_s$, then the element x_{ij} is placed in the r-th row and s-th column of the matrix W).
Now we can assign new values $sour_W[i]$ and $dest_W[j]$ $(1 \leq i \leq p, 1 \leq j \leq q)$ for matrix W:

$$sour_W[i] = \sum_{j \in \{j_1, j_2, \ldots, j_q\}} x_{ij}, \ 1 \leq i \leq p,$$
$$dest_W[j] = \sum_{i \in \{i_1, i_2, \ldots, i_p\}} x_{ij}, \ 1 \leq j \leq q.$$

We can initialize the matrix W (procedure *initialization*) so that all constraints $sour_W[i]$ and $dest_W[j]$ are satisfied. Then we replace corresponding elements of matrix V by new elements from the matrix W. In this way all the global constraints ($sour[i]$ and $dest[j]$) are preserved.

mutation-2. This operator is identical to mutation-1 except that in recalculating the contents of the chosen sub-matrix W, a modified version of the *initialization* routine is used (for details, see [36]) which avoids zero entries by selecting values from a range.

crossover. Starting with two parents (matrices U and V) the arithmetical crossover operator will produce two children X and Y, where $X = c_1 \cdot U + c_2 \cdot V$ and $Y = c_1 \cdot V + c_2 \cdot U$ (where $c_1, c_2 \geq 0$ and $c_1 + c_2 = 1$). As the constraint set is convex this operation ensures that both of children are feasible if both parents are. This is a significant simplification of the linear case where there was an additional requirement to maintain all components of the matrix as integers.

It is clear that all above operators maintain feasibility of potential solutions: (arithmetical) crossover produces a point between two feasible points of the convex search space and both mutations were restricted to submatrices only to ensure no change in marginal sums.

The experimental results of the developed system are discussed in [36, 34, 33]. It is worthwhile to underline, that the results were much better than these obtained from the GAMS (General Algebraic Modeling System)[5] with MINOS optimizer.

The main conclusions from the experiments on the transportation problem were as follows:

- the most interesting problems are constrainted ones,
- coding of chromosome structures S need not be binary,
- the 'genetic' operators need not be 'genetic' and may incorporate the problem specific knowledge, and
- the problem-specific knowledge incorporated into the system enhances an algorithm's performance and narrows its applicability.

In the next section we discuss further the last conclusion.

4 Evolutionary computation techniques and the problem specific knowledge

The idea of incorporating a problem specific knowledge in genetic algorithms is not new and has been recognized for some time. Several researchers have discussed initialization techniques, different representations, and the use of heuristics for genetic operators. In [10] Davis wrote:

---

[5] GAMS is a package for the construction and solution of mathematical programming models [8]. GAMS represents a typical example of an industry-standard efficient gradient-controlled method.

"It has seemed true to me for some time that we cannot handle most real-world problems with binary representations and an operator set consisting only of binary crossover and binary mutation. One reason for this is that nearly every real-world domain has associated domain knowledge that is of use when one is considering a transformation of a solution in the domain [...] I believe that genetic algorithms are the appropriate algorithms to use in a great many real-world applications. I also believe that one should incorporate real-world knowledge in one's algorithm by adding it to one's decoder or by expanding one's operator set."

The observation seems straightforward, however, it is not clear whether the resulting system should be still called 'genetic algorithm'. For arbitrary data structure S which implements the solution chromosome and expanded set of 'genetic' operators, we get just some evolutionary algorithm, which might be closer to other evolutionary techniques. The system discussed in the previous section is just one example of such phenomena.

In [33] we discussed efficiency versus problem spectrum for evolutionary computation methods. The main observation was that the problem-specific knowledge incorporated into the system enhances an algorithm's performance and narrows its applicability; it was summarized in Figure 2.

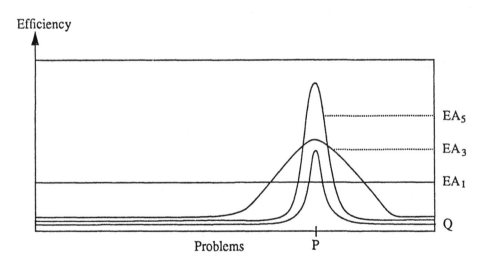

Fig. 2. Efficiency/problem spectrum and a hierarchy of EAs

The specialized algorithm Q is suitable for a particular problem P. In general, it is possible to construct a family of evolutionary algorithms EA_i, each of which would 'solve' the problem P. The term 'solve' means 'provide a reasonable solution', i.e., a solution which need not, of course, be optimal, but is feasible (it satisfies problem constraints). The evolutionary algorithm EA_5 is the strongest one (i.e., the most problem specific) and it addresses the problem P only. The system EA_5 will not work well for any modified version of the problem (e.g., after adding a new constraint or after changing the size of the problem). The next evolutionary algorithm, EA_4, can be applied to some (relatively small) class of problems, which includes the problem P; other evolutionary algorithms EA_3 and EA_2 work on larger domains, whereas EA_1 is the weakest method (i.e., domain independent) and can be applied to any optimization problem[6] (in [33] we discussed five such evolutionary algorithms for a given problem P: the transportation problem).

Note that we talk about evolutionary algorithms in general, without specifying whether such system is a genetic algorithm, an evolution strategy, an evolutionary programming technique, or other. The reason was explained partially in the previous section, where we discussed a particular evolutionary system for the transportation problem, which was hard to classify into these categories. It seems that neither of the evolutionary techniques is perfect (or even robust) across the problem spectrum; only the whole family of algorithms based on evolutionary computation concepts (i.e., evolutionary algorithms) have the property of robustness. For example, if we concentrate on classical GAs (binary representation, crossover and mutation) and take into account that [14]:

> "...virtually all decision making situations involve constraints. What distinguish various types of problems is the form of these constraints. Depending on how the problem is visualized, they can arise as rules, data dependencies, algebraic expressions, or other forms",

it is clear, that major modifications should be incorporated in the GA to obtain satisfactory results. Because of these modifications, we deal rather with some evolutionary algorithms (as discussed in the previous section). Classical GAs aim mainly at adaptive problems; their applications in the area of Artificial Life prove this point. Similary, evolution strategies aim at numerical optimization, whereas evolutionary programming—at task prediction.[7] In the same time, application of GAs to numerical optimization results often in systems more similar to evolution strategies than genetic algorithms [34]! In some sense we can look at the main three instances of evolutionary computation (i.e, genetic algorithms, evolution strategy, evolutionary programming) from the perspective of incorporating problem-specific knowledge into evolutionary algorithm. Each of these techniques incorporates the problem-specific knowledge in their data structures (binary string, floating point string representing values of variables and values of control parameters, finite state machines, respectively) and operators (binary crossover and mutation, Gaussian mutation, special mutations for finite state

---

[6] Note, that only EA_5, EA_3, and EA_1 are displayed in Figure 2.

[7] Only recently evolutionary programming techniques were extended to handle numerical optimization problems [16].

machines, respectively). This is why it is difficult (or even unfair) to compare these techniques on some particular subset of problems. Figure 3 illustrates these points.

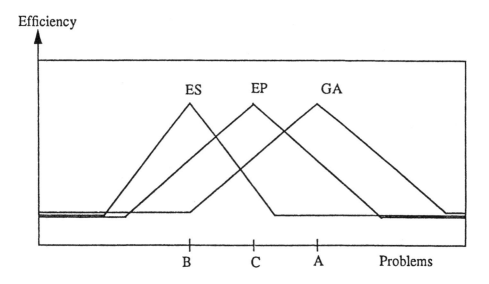

Fig. 3. Efficiency/problem spectrum and GA, ES, EP

In Figure 3 the problem A might be a problem of modeling ecological system, the problem B—numerical optimization problem, and C—for example, prisoner's dilemma problem.

In 1985 De Jong [12] addressed this issue from the perspective of genetic algorithms:

> "What should one do when elements in the space to be searched are most naturally represented by more complex data structures such as arrays, trees, digraphs, etc. Should one attempt to 'linearize' them into a string representation or are there ways to creatively redefine crossover and mutation to work directly on such structures."

It seems that more than often the second possibility (i.e., redefining crossover(s) and mutation(s) to work directly on complex structures) provides with better experimental results. This point of view is shared by Koza [31]:

> "Representation is the key issue in genetic algorithm work because the representation scheme can severely limit the window by which the system observes its world."

Koza developed a new methodology, named 'genetic programming' (GP), which provides a way to run a search of the space of possible computer programs for the best one (the most fit). In other words, a population of computer programs is created, individual programs compete against each other, weak programs die, and strong ones reproduce (crossover, mutation).... It is important to note that the structure which undergoes evolution is a hierarchically structured computer program.[8] The search space is a hyperspace of valid programs, which can be viewed as a space of rooted trees. Each tree is composed of functions and terminals appropriate to the particular problem domain; the set of all functions and terminals is selected *a priori* in such a way that some of the composed trees yield a solution. The initial population is composed of such trees; construction of a (random) tree is straightforward. The evaluation function assigns a fitness value which evaluates the performance of a tree (program). The evaluation is based on a preselected set of test cases; in general, the evaluation function returns the sum of distances between the correct and obtained results on all test cases. The primary operator is a crossover that produces two offspring from two selected parents. The crossover creates offspring by exchanging subtrees between two parents. There are other operators as well: mutation, permutation, editing, and a define-building-block operation [31]. The paradigm of genetic programming constitutes another important example of incorporating problem specific knowledge by means of data structures and operators used.

5 Current critical issues

There are several issues which deserve a special attention of the evolutionary computation community. As De Jong observed [13] recently:

> "... the field had pushed the application of simple GAs[9] well beyond our initial theories and understanding, creating a need to revisit and extend them."

In this section we examine a few important directions in which we expect a lot of activities and significant results.

Function optimization

For many years, most evolutionary techniques were evaluated and compared with each other in the domain of function optimization. In the view of the previous section, it is not surprising that quite often ES outperformed simple GAs; this was also the case with EP techniques when they were extended to handle numerical optimization problems [16]. To adapt a GA to the task of function optimization it was necessary to extend simple GA by additional features; these included

- dynamic scaled fitness,

---

[8] Actually, Koza has chosen LISP's S-expressions for all his experiments.

[9] This is true not only for GAs, but for any evolutionary technique.

- rank-proportional selection,
- inclusion of elitist strategy,
- adaptation of various parameters of the search (probabilities of operators, population size, etc.)
- various representations: binary or Gray coding (plus Delta Coding or Dynamic Parameter Encoding), and floating point representation,
- new operators (for binary and floating point representation).

Most of these modifications pushed a simple GA away from its theoretical basis, however, they enhanced the performance of the systems in a significant way. It seems that the domain of function optimization would remain as the primary test-bed for many new features. It is expected that new theories of evolutionary techniques for function optimization would emerge (e.g., breeder genetic algorithms [37]). Additionally, we should see a progress in

- development of constraint-handling techniques. This is a very important area in general, and for function optimization, in particular; most real problems of function optimization involve constraints. However, so far very few techniques were proposed, analysed, and compared with each other.
- development of systems for large-scale problems. Until now, most experiments assume relatively small number of variables. It would be interesting to analyse how evolutionary techniques scale up with the problem size for problems with thousand variables.
- development of systems for mathematical programming problems. Very little work was done in this area. There is a need to investigate evolutionary systems to handle integer/Boolean variables, to experiment with mixed programming as well as integer programming problems.

Representation and operators

Traditionally, GAs work with binary strings, ES—with floating point vectors, and EP—with finite state machines (represented as matrices), whereas GP techniques use trees as a structure for the individuals. However, there is a need for a systematic research on

- representation of complex, non-linear objects of varying size, and, in particular, representation of 'blueprints' of complex objects, and
- development of evolutionary operators for such objects at the genotype level.

This direction can be perceived as a step towards building complex hybrid evolutionary system which incorporate additional search techniques. For example, it seems worthwhile to experiment with Lamarckian operators, which would improve an individual during its lifetime—consequently, the improved, "learned" characteristics of such individual would be passed to the next generation.

Non random mating

Most current techniques which incorporate crossover operator use random mating, i.e, mating, where individuals are paired randomly. It seems that with the trend of movement from simple to complex systems, the issue of non random mating would be of growing importance. There are many possibilities to explore; these include introduction of sex or "family" relationships between individuals. Some simple schemes were already investigated by several researchers (e.g., Eshelman's incest prevention technique [15]), however, the ultimate goal seems to be to evolve rules for non random mating.

Self-adapting systems

Since evolutionary algorithms implement the idea of evolution, it is more than natural to expect some self-adapting characteristics of these techniques. Apart from evolutionary strategies, which incorporate some of its control parameters in the solution vectors, most other techniques use fixed representations, operators, and control parameters. One of the most promising research area is based on inclusion of self adapting mechanisms within the system for

- representation of individuals (as proposed by Shaefer [50]; the Dynamic Parameter Encoding technique [46] and messy genetic algorithms [25] also fall into this category),
- operators. It is clear that different operators play different roles at different stages of the evolutionary process. The operators should adopt (e.g., adaptive crossover [45]). This is true especially for time-varying fitness landscapes.
- control parameters. There were already experiments aimed at these issues: adaptive population sizes [2] or adaptive probabilities of operators [10]. However, much more remains to be done.

Co-evolutionary systems

There is a growing interest in co-evolutionary systems, where more than one evolution process takes place: usually there are different populations there (e.g., additional populations of parasites or predators) which interact with each other. In such systems the evaluation function for one population may depend on the state of the evolution processes in the other population(s). This is an important topic for modeling artificial life, some business applications, etc. Also, co-evolutionary systems might be important for approaching large-scale problems [40].

Diploid/polyploid versus haploid structures

Diploidy (or polyploidy) can be viewed as a way to incorporate memory into the individual's structure. Instead of single chromosome (haploid structure) representing a precise information about an individual, a diploid structure is made up of a pair of chromosomes: the choice between two values is made by some dominance function. The diploid (polyploid) structures are of particular significance in non-stationary environments (i.e., for time-varying objective functions) and for modeling complex systems (possibly using co-evolution models). However, there is no theory to support the incorporation of a dominance function into the system; there is also quite small experimental data in this area.

Parallel models

Parallelism promises to put within our reach solutions to problems untractable before; clearly, it is one of the most important areas of computer science. Evolutionary algorithms are very suitable for parallel implememtations; as Goldberg [24] observed:

"In a world where serial algorithms are usually made parallel through countless tricks and contortions, it is no small irony that genetic algorithms (highly parallel algorithms) are made serial through equally unnatural tricks and turns."

However, during the last 5 years several parallel models of evolutionary techniques were investigated. They can be classified into several categories (e.g., synchronous vs. asynchronous, or master-slave vs. network, fine-grain vs. coarse-grain, continuous vs. discontinuous neighborhood). Many experimental results indicate a significant speedup in the processing time, however, there is very little theory to assist in understanding parallel systems.

Parallel models can also provide a natural embedding for other paradigms of evolutionary computation, like non random mating, some aspects of self-adaptation, or co-evolutionary systems.

6 Conclusions

It might be meaningful to conclude the paper with a citation from the recent meeting of the evolutionary computation community, which provides the main direction for future research [22]:

"If the aim is to generate artificial intelligence, that is, to solve new problems in new ways, then it is inappropriate to use any fixed set of rules. The rules required for solving each problem should simply evolve..."

Acknowledgements.
This material is based upon work supported by the National Science Foundation under Grant IRI-9322400. The author wishes to thank Sita S. Raghavan for comments on the first draft of the paper.

References

1. Alander, J.T., *An Indexed Bibliography of Genetic Algorithms: Years 1957–1993*, Department of Information Technology and Production Economics, University of Vaasa, Finland, Report Series No.94-1, 1994.
2. Arabas, J., Michalewicz, Z., and Mulawka, J., *GAVaPS — a Genetic Algorithm with Varying Population Size*, in [41].
3. Bäck, T., Fogel, D., and Michalewicz, Z. (Editors), *Handbook of Evolutionary Computation*, IOP Press, in preparation.

4. Beasley, D., Bull, D.R., and Martin, R.R., *An Overview of Genetic Algorithms: Part 1, Foundations*, University Computing, Vol.15, No.2, pp.58–69, 1993.
5. Beasley, D., Bull, D.R., and Martin, R.R., *An Overview of Genetic Algorithms: Part 2, Research Topics*, University Computing, Vol.15, No.4, pp.170–181, 1993.
6. Belew, R. and Booker, L. (Editors), Proceedings of the Fourth International Conference on Genetic Algorithms, Morgan Kaufmann Publishers, Los Altos, CA, 1991.
7. De Jong, K.A., (Editor), *Evolutionary Computation*, MIT Press, 1993.
8. Brooke, A., Kendrick, D., and Meeraus, A., *GAMS: A User's Guide*, The Scientific Press, 1988.
9. Davis, L., (Editor), *Genetic Algorithms and Simulated Annealing*, Morgan Kaufmann Publishers, Los Altos, CA, 1987.
10. Davis, L., *Adapting Operator Probabilities in Genetic Algorithms*, in [44], pp.61–69.
11. Davis, L. and Steenstrup, M., *Genetic Algorithms and Simulated Annealing: An Overview*, in [9], pp.1–11.
12. De Jong, K., *Genetic Algorithms: A 10 Year Perspective*, in [27], pp.169–177.
13. De Jong, K., *Genetic Algorithms: A 25 Year Perspective*, in [53], pp.125–134.
14. Dhar, V. and Ranganathan, N., *Integer Programming vs. Expert Systems: An Experimental Comparison*, Communications of ACM, Vol.33, No.3, pp.323–336, 1990.
15. Eshelman, L.J. and Schaffer, J.D., *Preventing Premature Convergence in Genetic Algorithms by Preventing Incest*, in [6], pp.115–122.
16. Fogel, D.B., *Evolving Artificial Intelligence*, Ph.D. Thesis, University of California, San Diego, 1992.
17. Fogel, D.B. (Editor), IEEE Transactions on Neural Networks, special issue on Evolutionary Computation, Vol.5, No.1, 1994.
18. Fogel, D.B., *An Introduction to Simulated Evolutionary Optimization*, IEEE Transactions on Neural Networks, special issue on Evolutionary Computation, Vol.5, No.1, 1994.
19. Fogel, D.B. and Atmar, W., *Proceedings of the First Annual Conference on Evolutionary Programming*, La Jolla, CA, 1992, Evolutionary Programming Society.
20. Fogel, D.B. and Atmar, W., *Proceedings of the Second Annual Conference on Evolutionary Programming*, La Jolla, CA, 1993, Evolutionary Programming Society.
21. Fogel, L.J., Owens, A.J., and Walsh, M.J., *Artificial Intelligence Through Simulated Evolution*, John Wiley, Chichester, UK, 1966.
22. Fogel, L.J., *Evolutionary Programming in Perspective: The Top-Down View*, in [53], pp.135–146.
23. Forrest, S. (Editor), Proceedings of the Fifth International Conference on Genetic Algorithms, Morgan Kaufmann Publishers, Los Altos, CA, 1993.
24. Goldberg, D.E., *Genetic Algorithms in Search, Optimization and Machine Learning*, Addison-Wesley, Reading, MA, 1989.
25. Goldberg, D.E., Deb, K., and Korb, B., *Do not Worry, Be Messy*, in [6], pp.24–30.
26. Goldberg, D.E., Milman, K., and Tidd, C., *Genetic Algorithms: A Bibliography*, IlliGAL Technical Report 92008, 1992.
27. Grefenstette, J.J., (Editor), Proceedings of the First International Conference on Genetic Algorithms, Lawrence Erlbaum Associates, Hillsdale, NJ, 1985.
28. Grefenstette, J.J., (Editor), Proceedings of the Second International Conference on Genetic Algorithms, Lawrence Erlbaum Associates, Hillsdale, NJ, 1987.
29. Heitkötter, J., (Editor), *The Hitch-Hiker's Guide to Evolutionary Computation*, FAQ in comp.ai.genetic, issue 1.10, 20 December 1993.
30. Holland, J.H., *Adaptation in Natural and Artificial Systems*, University of Michigan Press, Ann Arbor, 1975.

31. Koza, J., *Genetic Programming*, MIT Press, 1992.

32. Männer, R. and Manderick, B. (Editors), Proceedings of the Second International Conference on Parallel Problem Solving from Nature (PPSN), North-Holland, Elsevier Science Publishers, Amsterdam, 1992.

33. Michalewicz, Z., *A Hierarchy of Evolution Programs: An Experimental Study*, Evolutionary Computation, Vol.1, No.1, 1993, pp.51–76.

34. Michalewicz, Z., *Genetic Algorithms + Data Structures = Evolution Programs*, Springer-Verlag, 2nd edition, 1994.

35. Michalewicz, Z. (Editor), Statistics & Computing, special issue on evolutionary computation, Vol.4, No.2, 1994.

36. Michalewicz, Z., Vignaux, G.A., and Hobbs, M., *A Non-Standard Genetic Algorithm for the Nonlinear Transportation Problem*, ORSA Journal on Computing, Vol.3, No.4, 1991, pp.307–316.

37. Mühlenbein, H. and Schlierkamp-Vosen, D., *Predictive Models for the Breeder Genetic Algorithm*, Evolutionary Computation, Vol.1, No.1, pp.25–49, 1993.

38. Nissen, V., *Evolutionary Algorithms in Management Science: An Overview and List of References*, European Study Group for Evolutionary Economics, 1993.

39. Orvosh, D. and Davis, L., *Shall We Repair? Genetic Algorithms, Combinatorial Optimization, and Feasibility Constraints*, in [23], p.650.

40. Potter, M. and De Jong, K., *A Cooperative Coevolutionary Approach to Function Optimization*, George Mason University, 1994.

41. Proceedings of the First IEEE International Conference on Evolutionary Computation, Z. Michalewicz, J.D. Schaffer, H.-P. Schwefel, H. Kitano, D. Fogel (Editors), Orlando, 26 June – 2 July, 1994.

42. Reeves, C.R., *Modern Heuristic Techniques for Combinatorial Problems*, Blackwell Scientific Publications, London, 1993.

43. Saravanan, N. and Fogel, D.B., *A Bibliography of Evolutionary Computation & Applications*, Department of Mechanical Engineering, Florida Atlantic University, Technical Report No. FAU-ME-93-100, 1993.

44. Schaffer, J., (Editor), Proceedings of the Third International Conference on Genetic Algorithms, Morgan Kaufmann Publishers, Los Altos, CA, 1989.

45. Schaffer, J.D. and Morishima, A., *An Adaptive Crossover Distribution Mechanism for Genetic Algorithms*, in [28], pp.36–40.

46. Schraudolph, N. and Belew, R., *Dynamic Parameter Encoding for Genetic Algorithms*, CSE Technical Report #CS90-175, University of San Diego, La Jolla, 1990.

47. Schwefel, H.-P., *On the Evolution of Evolutionary Computation*, in [53], pp.116–124.

48. Schwefel, H.-P. and Männer, R. (Editors), Proceedings of the First International Conference on Parallel Problem Solving from Nature (PPSN), Springer-Verlag, Lecture Notes in Computer Science, Vol.496, 1991.

49. Sebald, A.V. and Fogel, L.J., *Proceedings of the Third Annual Conference on Evolutionary Programming*, San Diego, CA, 1994, World Scientific.

50. Shaefer, C.G., *The ARGOT Strategy: Adaptive Representation Genetic Optimizer Technique*, in [28], pp.50–55.

51. Vignaux, G.A., and Michalewicz, Z., *A Genetic Algorithm for the Linear Transportation Problem*, IEEE Transactions on Systems, Man, and Cybernetics, Vol.21, No.2, 1991, pp.445–452.

52. Whitley, D., *Genetic Algorithms: A Tutorial*, in [35], pp.65–85.

53. Zurada, J., Marks, R., and Robinson, C. (Editors), *Computational Intelligence: Imitating Life*, IEEE Press, 1994.

An Experimental Study of N-Person Iterated Prisoner's Dilemma Games

Xin Yao and Paul J. Darwen

Department of Computer Science
University College, The University of New South Wales
Australian Defence Force Academy
Canberra, ACT, Australia 2600

Abstract. The Iterated Prisoner's Dilemma game has been used extensively in the study of the evolution of cooperative behaviours in social and biological systems. There have been a lot of experimental studies on evolving strategies for 2-player Iterated Prisoner's Dilemma games (2IPD). However, there are many real world problems, especially many social and economic ones, which cannot be modelled by the 2IPD. The n-player Iterated Prisoner's Dilemma (NIPD) is a more realistic and general game which can model those problems. This paper presents two sets of experiments on evolving strategies for the NIPD. The first set of experiments examine the impact of the number of players in the NIPD on the evolution of cooperation in the group. Our experiments show that cooperation is less likely to emerge in a large group than in a small group. The second set of experiments study the generalisation ability of evolved strategies from the point of view of machine learning. Our experiments reveal the effect of changing the evolutionary environment of evolution on the generalisation ability of evolved strategies.

1 Introduction

The 2-player Iterated Prisoner's Dilemma game (2IPD) is a 2×2 non-zerosum noncooperative game, where "non-zerosum" indicates that the benefits obtained by a player are not necessarily the same as the penalties received by another player and "noncooperative" indicates that no preplay communication is permitted between the players [1, 2]. It has been widely studied in such diverse fields as economics, mathematical game theory, political science, and artificial intelligence.

In the Prisoner's Dilemma, each player has a choice of two operations: either cooperate with the other player, or defect. Payoff to both players is calculated according to Figure 1. In the Iterated Prisoner's Dilemma (IPD), this step is repeated many times, and each player can remember previous steps.

While the 2IPD has been studied extensively for more than three decades, there are many real world problems, especially many social and economic ones, which cannot be modelled by the 2IPD. Hardin [3] described some examples of such problems. More examples can be found in Colman's book [1](pp.156–159). The n-player Iterated Prisoner's Dilemma (NIPD) is a more realistic and general

	Cooperate	Defect
Cooperate	R R	T S
Defect	S T	P P

Fig. 1. The payoff matrix for the 2-player prisoner's dilemma game. The values S, P, R, T must satisfy $T > R > P > S$ and $R > (S + T)/2$. In 2-player Iterated Prisoner's Dilemma (2IPD), the above interaction is repeated many times, and both players can remember previous outcomes.

game which can model those problems. In comparing the NIPD with the 2IPD, Davis *et al.* [4](pp.520) commented that

> The N-player case (NPD) has greater generality and applicability to real-life situations. In addition to the problems of energy conservation, ecology, and overpopulation, many other real-life problems can be represented by the NPD paradigm.

Colman [1](pp.142) and Glance and Huberman [5, 6] have also indicated that the NIPD is "qualitatively different" from the 2IPD and that "... certain strategies that work well for individuals in the Prisoner's Dilemma fail in large groups."

The n-player Prisoner's Dilemma game can be defined by the following three properties [1](pp.159):

1. each player faces two choices between cooperation (C) and defection (D);
2. the D option is dominant for each player, i.e., each is better off choosing D than C no matter how many of the other players choose C;
3. the dominant D strategies intersect in a deficient equilibrium. In particular, the outcome if all players choose their non-dominant C strategies is preferable from every player's point of view to the one in which everyone chooses D, but no one is motivated to deviate unilaterally from D.

Figure 2 shows the payoff matrix of the n-player game.

A large number of values satisfy the requirements of Figure 2. We choose values so that, if n_c is the number of cooperators in the n-player game, then the payoff for cooperation is $2n_c - 2$ and the payoff for defection is $2n_c + 1$. Figure 3 shows an example of the n-player game.

With this choice, simple algebra reveals that if N_c cooperative moves are made out of N moves of an n-player game, then the average per-round payoff a is given by:

Number of cooperators among the remaining $n - 1$ players

		0	1	2		$n - 1$
	C	C_0	C_1	C_2	$\cdots$	C_{n-1}
player A						
	D	D_0	D_1	D_2	$\cdots$	D_{n-1}

Fig. 2. The payoff matrix of the n-player Prisoner's Dilemma game, where the following conditions must be satisfied: (1) $D_i > C_i$ for $0 \le i \le n - 1$; (2) $D_{i+1} > D_i$ and $C_{i+1} > C_i$ for $0 \le i < n - 1$; (3) $C_i > (D_i + C_{i-1})/2$ for $0 < i \le n - 1$. The payoff matrix is symmetric for each player.

Number of cooperators among the remaining $n - 1$ players

		0	1	2		$n - 1$
	C	0	2	4	$\cdots$	$2(n - 1)$
player A						
	D	1	3	5	$\cdots$	$2(n - 1) + 1$

Fig. 3. An example of the N-player game.

$$a = 1 + \frac{N_c}{N}(2n - 3) \tag{1}$$

This lets us measure how common cooperation was just by looking at the average per-round payoff.

There has been a lot of research on the evolution of cooperation in the 2IPD using genetic algorithms and evolutionary programming in recent years [7, 8, 9, 10, 11, 12]. Axelrod [7] used genetic algorithms to evolve a population of strategies where each strategy plays the 2IPD with every other strategy in the population. In other words, the performance or fitness of a strategy is evaluated by playing the 2IPD with every other strategy in the population. The environment in which a strategy evolves consists of all the remaining strategies in the population. Since strategies in the population are constantly changing as a result of evolution, a strategy will be evaluated by a different environment in every generation. All the strategies in the population are co-evolving in their dynamic environments. Axelrod found that such dynamic environments produced strategies that performed very well against their population. Fogel [11] described similar experiments, but used finite state machines to represent strategies and evolutionary programming to evolve them.

However, very few experimental studies have been carried out on the NIPD in spite of its importance and its qualitative difference from the 2IPD. This paper presents two sets of experiments carried out on the NIPD. We first describe our experiment setup in Section 2. Then we investigate the impact of the number of players in the Prisoner's Dilemma game on the evolution of cooperation in Section 3. We are mainly interested in two questions here: (1) whether cooperation can still emerge from a larger group, and (2) whether it is more difficult to evolve cooperation in a larger group. The evolution of strategies for the NIPD can be regarded as a form of machine learning using the evolutionary approach. An important issue in machine learning is generalisation. Section 4 of this paper discusses the generalisation issue associated with co-evolutionary learning and presents some experiments with different evolutionary environments. Finally, Section 5 concludes with some remarks and future research directions.

2 Experiment Setup

2.1 Genotypical Representation of Strategies

We use genetic algorithms to evolve strategies for the NIPD. The most important issue here is the representation of strategies. We will use two different representations, both of which are look-up tables that give an action for every possible contingency.

One way of representing strategies for the NIPD is to generalise the representation scheme used by Axelrod [7]. In this scheme, each genotype is a lookup table that covers every possible history of the last few steps. A history in such a game is represented as a binary string of ln bits, where the first l bits represent the player's own previous l actions (most recent to the left, oldest to the right), and the other $n-1$ groups of l bits represent the previous actions of the other players. For example, during a game of 3IPD with a remembered history of 2 steps, $n = 3, l = 2$, one player might see this history:

$$n = 3, l = 2: \text{Example history } 11 \ 00 \ 01$$

The first l bits, 11, means this player has defected (a "1") for both of the previous $l = 2$ steps. The previous steps of the other players are then listed in order: the 00 means the first of the other players cooperated (a "0") on the previous l steps, and the last of the other players cooperated (0) on the most recent step, and defected (1) on the step before, as represented by 01.

For the NIPD remembering l previous steps, there are 2^{ln} possible histories. The lookup table genotype therefore contains an action (cooperate "0" or defect "1") for each of these possible histories. So we need at least 2^{ln} bits to represent a strategy. At the beginning of each game, there are no previous l steps of play from which to look up the next action, so each genotype should also contain its own extra bits that define the presumed pre-game moves. The total genotype length is therefore $2^{ln} + ln$ bits. We will use this genotype for the first set of results below, Figure 5 through to Figure 8.

This Axelrod-style representation scheme, however, suffers from two disadvantages. First, it does not scale well as the number of players increases. Second, it provides more information than is necessary by telling which of the other players cooperated or defected, when the only information needed is how many of the other players cooperated or defected. Such redundant information had reduced the efficiency of the evolution greatly in our experiments with this representation scheme. To improve on this, we use a new representation scheme which is more compact and efficient.

In our new representation scheme, each individual is regarded as a set of rules stored in a look-up table that covers every possible history. As a game that runs for, say, 500 rounds would have an enormous number of possible histories, and as only the most recent steps will have significance for the next move, we only consider every possible history over the most recent l steps, where l is less than 4 steps. This means an individual can only remember the l most recent rounds. Such a history of l rounds is represented by:

1. l bits for the player's own previous l moves, where a "1" indicates defection, a "0" cooperation; and
2. another $l \log_2 n$ bits for the number of cooperators among the other $n - 1$ players, where n is the number of the players in the game. This requires that n is a power of 2.

For example, if we are looking at 8 players who can remember the 3 most recent rounds, then one of the players would see the history as:

History for 8 players, 3 steps: 001 111 110 101 (12 bits)

Here, the 001 indicates the player's own actions: the most recent action (on the left) was a "0", indicating cooperation, and the action 3 steps ago (on the right), was a "1", i.e., defection. The 111 gives the number of cooperators among the other 7 players in the most recent round, i.e., there were $111_2 = 7$ cooperators. The 101 gives the number of cooperators among the other 7 players 3 steps ago, i.e., there were $101_2 = 5$ cooperators. The most recent events are always on the left, previous events on the right.

In the above example, there are $2^{12} = 2048$ possible histories. So 2048 bits are needed to represent all possible strategies. In the general case of an n-player game with history length l, each history needs $l + l \log_2 n$ bits to represent and there are $2^{l + l \log_2 n}$ such histories. A strategy is represented by a binary string that gives an action for each of those possible histories. In the above example, the history 001 111 110 101 would cause the strategy to do whatever is listed in bit 1013, the decimal number for the binary 001111110101.

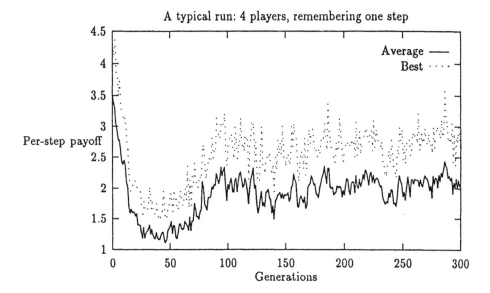

Fig. 4. This shows the average and best payoff at each generation for a population of 100 individuals. Each individual is a strategy.

Since there are no previous l rounds at the beginning of a game, we have to specify them with another $l(1 + \log_2 n)$ bits. Hence each strategy is finally represented by a binary string of length $2^{l+l \log_2 n} + l(1 + \log_2 n)$.

2.2 Genetic Algorithm Parameters

For all the experiments presented in this paper, the population size is 100, the mutation rate is 0.001, and the crossover rate is 0.6. Rank-based selection was used, with the worst performer assigned an average of 0.75 offspring, the best 1.25 offspring.

2.3 A Typical Run

A tyical run with four players with a history 1 $(n = 4, l = 1)$ is shown in Figure 4. At each generation, 1000 games of the 4-player Iterated Prisoner's Dilemma are played, with each group of 4 players selected randomly with replacement. Each of these 1000 games lasts for 100 rounds. Starting from a random population, defection is usually the better strategy, and the average payoff plummets initially. As time passes, some cooperation becomes more profitable. We will examine more results in detail later.

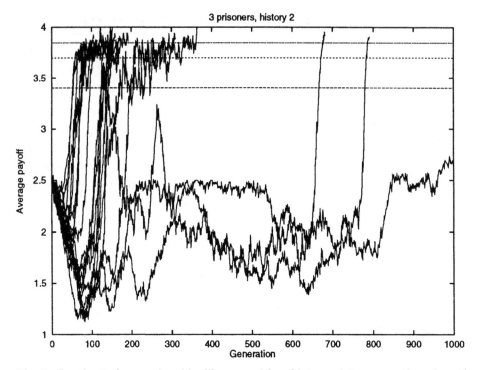

Fig. 5. For the 3-player prisoner's dilemma with a history of 2, cooperation almost always emerges. Only 1 out of 20 runs fail to reach 95% cooperation using Axelrod's representation shceme.

3 Group Size of the NIPD

This section discusses the impact of group size, i.e., the number of players in the NIPD, on the evolution of cooperation and presents some experimental results. It is well-known that cooperation can be evolved from a population of random strategies for the 2IPD. Can cooperation still be evolved from a population of strategies for the NIPD where the number of players is greater than 2? If the answer is yes, does the group size affect the evolution of cooperation in the NIPD?

Using the Axelrod-style genotype described above, we carried out a series of experiments with the 3IPD, 4IPD, 5IPD, and 6IPD games. In each of the following runs, the program stopped when more than 5 generations passed with the average payoff above the 95% cooperation level. Figure 5 shows the results of 20 runs of the 3IPD game with history length 2: out of 20 runs, there is only 1 which fails to reach 95% cooperation. Figure 6 shows the results of 20 runs of the 4IPD game with history length 2: 4 out of 20 runs fail to reach the 95% cooperation level, but only 1 of those fails to reach 80% cooperation. Figure 7 shows the results of 20 runs of the 5IPD game with history length 2: 6 out of 20 runs do not reach the 80% cooperation level. Figure 8 shows the results of 20

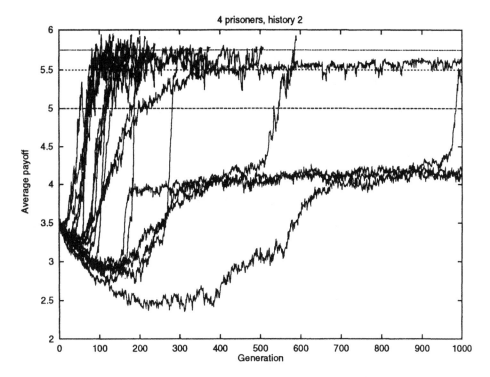

Fig. 6. For the 4-player prisoner's dilemma with a history of 2, cooperation almost always emerges. Only 4 out of 20 runs fail to reach 95% cooperation using Axelrod's representation shceme.

runs of the 6IPD game with history length 2: 9 out of 20 runs stay below the 80% cooperation level.

Figures 5 through 8 demonstrate that the evolution of cooperation becomes less likely as group size increases. Nonetheless, cooperation still emerges most of the time. As Axelrod's representation scheme used in those figures does not scale well with the group size, we use the second representation scheme described in Section 2 to carry out experiments with larger groups.

We have carried out a series of experiments with the 2IPD, 4IPD, 8IPD, and 16IPD games. Figure 9 shows the results of 10 runs of the 2IPD game with history length 3. Out of 10 runs, there are only 3 which fail to reach 90% cooperation and only 1 which goes to almost all defection. Figure 10 shows the results of 10 runs of the 4IPD game with history length 3, where some of the runs reach cooperation but more than half of the 10 runs fail to evolve cooperation. Figure 11 shows the results of 10 runs of the 8IPD game with history length 2, where none of the runs reach cooperation. Figure 12 shows the population bias in the runs in Figure 11, to demonstrate that those populations have pretty much converged. Figure 13 shows 10 runs of the 16IPD game.

These results confirm that cooperation can still be evolved in larger groups,

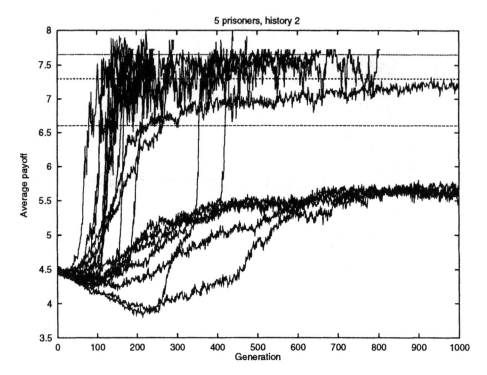

Fig. 7. For the 5-player prisoner's dilemma with a history of 2, cooperation almost always emerges. 6 out of 20 runs fail to reach 80% cooperation using Axelrod's representation shceme.

but it is more difficult to evolve cooperation as the group size increases. Glance and Huberman [5, 6] have arrived at a similar conclusion using a model based on many particle systems. We first suspected that the failure to evolve cooperation in larger groups was caused by larger search spaces and insufficient running time since more players were involved in 8IPD and 16IPD games. This is, however, not the case. The search space of the 8IPD game with history length 2 is actually smaller than that of the 4IPD game with history length 3. To confirm that the failure to evolve cooperation is not caused by insufficient running time, we examined the convergence of the 8IPD game. Figure 12 shows that at generation 200 the population has mostly converged for all the 10 runs.

It is worth mentioning that the evolution of cooperation using simulations does depend on some implementation details, such as the genotypical representation of strategies and the values used in the payoff matrix. So cooperation may be evolved in the 8IPD game if a different representation scheme and different payoff values are used. Although we cannot prove it vigorously, we think for any representation scheme and payoff values there would always be an upper limit on the group size over which cooperation cannot be evolved. Our experimental finding is rather similar to some phenomena in our human society, e.g., cooperation is usually easier to emerge in a small group of people than in a larger one.

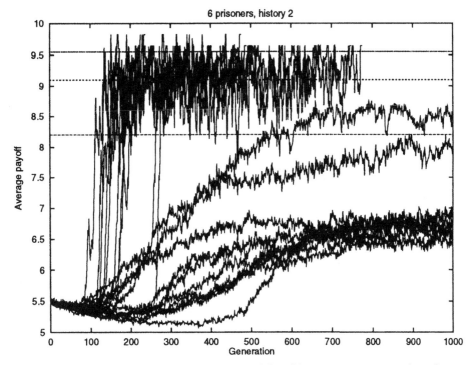

Fig. 8. For the 6-player prisoner's dilemma with a history of 2, cooperation almost always emerges. 9 1 out of 20 runs fail to reach 80% cooperation using Axelrod's representation shceme.

4 Co-Evolutionary Learning and Generalisation

The idea of having a computer algorithm learn from its own experience and thus create expertise without being exposed to a human teacher has been around for a long time. For genetic algorithms, both Hillis [13] and Axelrod [7] have attempted co-evolution, where a GA population is evaluated by how well it performs against itself or another GA population, starting from a random population. Expertise is thus bootstrapped from nothing, without an expert teacher. This is certainly an promising idea, but does it work? So far, no-one has investigated if the results of co-evolutionary learning are robust, that is, whether they generalise well? If a strategy is produced by a co-evolving population, will that strategy perform well against opponents never seen by that population? In order to investigate this issue, we need to pick the best strategies produced by the co-evolutionary learning system and let them play against a set of test strategies which had not been seen by the co-evolutionary system. This section describes some experiments which test the generalisation ability of co-evolved strategies for the 8IPD game with history length 1.

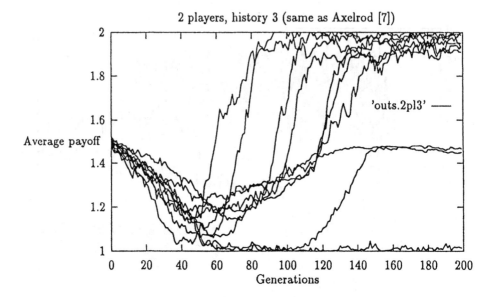

Fig. 9. For 2-player prisoner's dilemma with a history of 3, cooperation emerges most of the time. Only 3 out of 10 runs fail to reach 90% cooperation, and only 1 run goes to almost all defection.

4.1 Test Strategies

The unseen test strategies used in our study should be of reasonable standard and representative, that is, they are neither very poor (or else they will be exploited by their evolved opponents) nor very good (or else the will exploit their evolved opponent). We need unseen strategies that are adequate against a large range of opponents, but not *the* best.

To obtain such strategies, we did a limited enumerative search to find the strategies that performed best against a large number of random opponents. As most random opponents are very stupid, beating many random opponents provides a mediocre standard of play against a wide range of opponents. We limited this search to manageable proportions by fixing certain bits in a strategy's genotype that seemed to be sensible, such as always defecting after every other strategy defects. The top few strategies found from such a limited enumerative search are listed in Table 1.

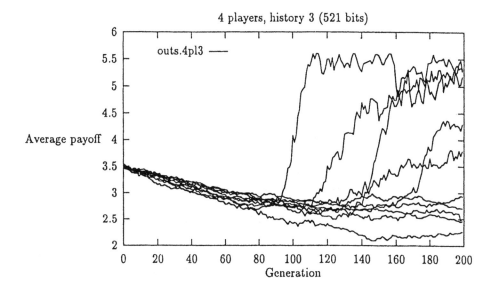

Fig. 10. For 10 runs of 4-player prisoner's dilemma with a history of 3, cooperation breaks out some of the time.

4.2 Learning and Testing

We have compared three different methods for implementing the co-evolutionary learning system. The three methods differ in the way each individual is evaluated, i.e., which opponents are chosen to evaluate an individual's fitness. The three methods are

1. Choosing from among the individuals in the GA population, i.e., normal co-evolution of a single population like Axelrod's implementation [7];
2. Choosing from a pool made of the evolving GA population and the best 25 strategies from the enumerative search, which remain fixed;
3. Choosing from a pool made of the evolving GA population and the best 25 strategies from the enumerative search, but the probability of choosing one of the 25 is four times higher.

For each of these, we obtained the best 25 strategies from the last generation of the GA, and tested it against a pool made up of both the seen and unseen enumerative search strategies, 50 in all.

4.3 Experimental Results

For each of the three evaluation methods, Tables 2 through 4 show the performance of the best strategies from the GA's last generation against opponents

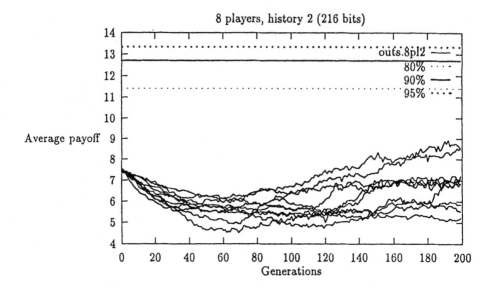

Fig. 11. For 10 runs of 8-player prisoner's dilemma with a history of 2, cooperation never emerges. The horizontal lines at the top show the 95%, 90%, and 80% levels of cooperation. To demonstrate that these runs have converged, figure 12 shows the bias of the populations.

Mean	Std Dev	Decimal	Binary genotype				
8.100	0.083	1026040	1111	1010	0111	1111	1000
8.093	0.083	1022965	1111	1001	1011	1111	0101
8.091	0.083	1018871	1111	1000	1011	1111	0111
8.088	0.083	1032181	1111	1011	1111	1111	0101
8.088	0.083	1020921	1111	1001	0011	1111	1001
8.082	0.083	1028087	1111	1010	1111	1111	0111
8.077	0.083	1023990	1111	1001	1111	1111	0110
8.076	0.083	1037305	1111	1101	0011	1111	1001
8.076	0.083	1017846	1111	1000	0111	1111	0110

Table 1. Top few strategies from a partial enumerative search for strategies that play well against a large number of random opponents. This provides unseen test opponents to test the generalisation of strategies produced by co-evolution. The first 4 bits were fixed to "1", as were the eleventh through sixteenth bits. Virtually all of the best 50 strategies started by cooperating.

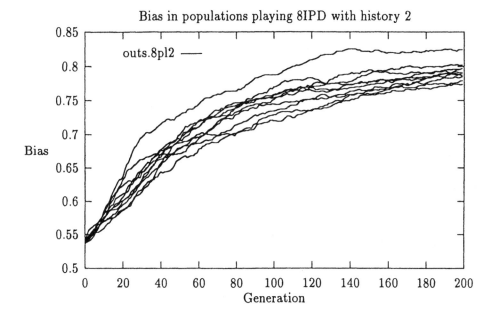

Fig. 12. In 10 runs of 8-player prisoner's dilemma with a history of 2, where cooperation never emerges, the bias demonstrates that the populations have converged. Bias is the average proportion of the most prominent value in each position. A bias of 0.75 means that, on average, each bit position has converged to either 75% "0" or 75% "1".

from (1) themselves, and (2) a pool made up of both the seen and unseen strategies from the enumerative search.

4.4 Discussion

Table 2 demonstrates that the co-evolution with the 8IPD produces strategies that are not very cooperative, as also demonstrated in Figure 11 earlier. Since the 8IPD is a game where it is easy to get exploited, co-evolution will first create strategies that can deal with non-cooperative strategies. The evolved strategies in Table 2 are cautious with each other and are not exploited by the unseen strategies from the enumerative search.

Adding fixed but not very cooperative strategies to the GA's evaluation procedure has a surprising effect. The evolved strategies in Tables 3 and 4 can cooperate well with other cooperators without being exploited by the strategies from the enumerative search, half of which it has never seen before. That is, normal co-evolution produces strategies which don't cooperate well with each other, and are not exploited by unseen non-cooperative strategies. Co-evolution with the addition of extra non-cooperative strategies gives more general strategies that do cooperate well with each other, but are still not exploited by unseen non-cooperative strategies. The experimental results also seem to indicate that the evolved strategies learn to cooperate with other cooperators better while maintaining their ability in dealing with non-cooperative strategies when the evolutionary environment contains a higher proportion of extra fixed strategies.

5 Conclusion

This paper describes two sets of experiments on the NIPD. The first set of experiments on the group size of the NIPD demonstrate that cooperation can still be evolved in the n-player IPD game where $n > 2$. However, it is more difficult to evolve cooperation as the group size increases. There are two research issues here which are worth pursuing; one is the upper limit of the group size over which cooperation cannot be evolved, the other is the quantitative relation between the group size and the time used to evolve cooperation. Glance and Huberman [5, 6] have addressed these two issues, but did not give a complete answer.

The second set of experiments in this paper deals with an important issue in co-evolutionary learning — the generalisation issue. Although the issue is the main theme in machine learning, very few people in the evolutionary computation community seem to be interested in it or address the issue explicitly and directly. We have presented some experimental results which show the importance of the environments in which each individual is evaluated, and their effects on generalisation ability.

References

1. A. M. Colman, *Game Theory and Experimental Games*, Pergamon Press, Oxford, England, 1982.
2. A. Rapoport, Optimal policies for the prisoner's dilemma, Technical Report 50, The Psychometric Lab., Univ. of North Carolina, Chapel Hill, NC, USA, July 1966.
3. G. Hardin, The tragedy of the commons, *Science*, 162:1243–1248, 1968.
4. J. H. Davis, P. R. Laughlin, and S. S. Komorita, The social psychology of small groups, *Annual Review of Psychology*, 27:501–542, 1976.
5. N. S. Glance and B. A. Huberman, The outbreak of cooperation, *Journal of Mathematical Sociology*, 17(4):281–302, 1993.
6. N. S. Glance and B. A. Huberman, The dynamics of social dilemmas, *Scientific American*, pages 58–63, March 1994.
7. R. Axelrod, The evolution of strategies in the iterated prisoner's dilemma, In L. Davis, editor, *Genetic Algorithms and Simulated Annealing*, chapter 3, pages 32–41. Morgan Kaufmann, San Mateo, CA, 1987.
8. D. M. Chess, Simulating the evolution of behaviors: the iterated prisoners' dilemma problem, *Complex Systems*, 2:663–670, 1988.
9. K. Lindgren, Evolutionary phenomena in simple dynamics, In C. G. Langton, C. Taylor, J. D. Farmer, and S. Rasmussen, editors, *Artificial Life II: SFI Studies in the Sciences of Complexity, Vol. X*, pages 295–312, Reading, MA, 1991. Addison-Wesley.

10. D. B. Fogel, The evolution of intelligent decision making in gaming, *Cybernetics and Systems: An International Journal*, 22:223–236, 1991.

11. D. B. Fogel, Evolving behaviors in the iterated prisoner's dilemma, *Evolutionary Computation*, 1(1):77–97, 1993.

12. P. J. Darwen and X. Yao, On evolving robust strategies for iterated prisoner's dilemma, In X. Yao, editor, *Proc. of the AI'93 Workshop on Evolutionary Computation*, pages 49–63, Canberra, Australia, November 1993. University College, UNSW, Australian Defence Force Academy.

13. W. Daniel Hillis, Co-evolving parasites improve simulated evolution as an optimization procedure, In *Santa Fe Institute Studies in the Sciences of Complexity, Volume 10*, pages 313–323. Addison-Wesley, 1991.

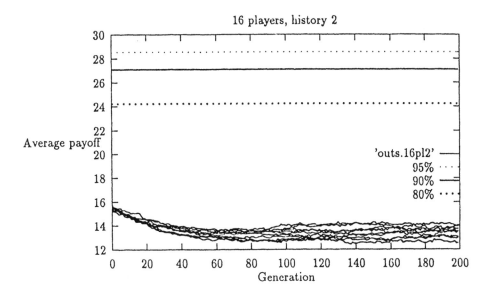

Fig. 13. For 10 runs of 16-player prisoner's dilemma with a history of 2, cooperation never emerges. The horizontal lines at the top show the 95%, 90%, and 80% levels of cooperation.

Normal co-evolution, no extra strategies in evaluation.
GA strategies play against themselves.

	Mean	Stdv	Stdv of mean	Mean of opponents

8pl1 (35% cooperative) against itself

		Mean	Stdv	Stdv of mean	Mean of opponents
0	11100001111011010011	7.240	3.978	0.126	6.296
1	11000001111011011011	7.285	3.985	0.126	6.392
2	11100000111111110011	7.335	3.052	0.097	8.434
3	01100010111111010011	7.258	3.202	0.101	7.889
4	11100101111111110011	7.180	4.160	0.132	5.883
5	01100000111111010011	7.000	3.090	0.098	8.111
6	11100101111111010011	7.171	4.125	0.130	6.127
7	01100001111111010011	7.241	4.027	0.127	6.286
8	00100000111111010110	7.165	3.122	0.099	8.341
9	01100100111111110110	7.706	3.523	0.111	7.949
10	10100000111111110001	7.274	3.083	0.097	8.395

GA strategies play against unseen strategies from enumerative search.

		Mean	Stdv	Stdv of mean	Mean of opponents
0	11100001111011010011	5.525	2.330	0.074	5.340
1	11000001111011011011	5.627	2.421	0.077	5.502
2	11100000111111110011	5.605	2.568	0.081	5.027
3	01100010111111010011	5.087	2.064	0.065	5.419
4	11100101111111110011	5.283	2.210	0.070	4.532
5	01100000111111010011	5.477	2.547	0.081	6.337
6	11100101111111010011	5.116	1.877	0.059	4.473
7	01100001111111010011	5.392	2.370	0.075	5.378
8	00100000111111010110	5.385	2.531	0.080	6.530
9	01100100111111110110	5.146	2.271	0.072	5.237
10	10100000111111110001	5.461	2.383	0.075	4.900

Table 2. Results of ordinary co-evolution, with no extra strategies during the GA evaluation. The GA strategies manage some cooperation among themselves, and hold their own against strategies they have not seen before.

Co-evolution, with addition of 25 fixed strategies from enumerative search.
GA strategies play against themselves.

		Mean	Stdv	Stdv of mean	Mean of opponents
0	11111000011111110100	11.678	1.715	0.054	11.965
1	11111000011111110100	11.706	1.553	0.049	11.994
2	11111000011111110110	11.440	1.603	0.051	11.922
3	11111000011111111110	11.721	1.581	0.050	12.027
4	11111000111111110100	13.264	2.521	0.080	10.636
5	11111000011111111110	11.714	1.584	0.050	12.025
6	11111000011111110110	11.420	1.669	0.053	11.895
7	11111000011111110100	11.678	1.705	0.054	11.985
8	11011000001111110100	11.618	1.781	0.056	11.958
9	11111000011111111111	11.670	1.688	0.053	11.974
10	11111000011111110100	11.649	1.697	0.054	11.973

GA strategies play against pool of 25 seen and 25 unseen strategies from enumerative search.

		Mean	Stdev	Stddev of mean	Mean of opponents
0	11111000011111110100	5.209	3.212	0.102	5.634
1	11111000011111110100	5.494	3.451	0.109	5.828
2	11111000011111110110	5.152	2.771	0.088	5.934
3	11111000011111111110	5.600	3.561	0.113	5.907
4	11111000111111110100	5.619	2.929	0.093	4.629
5	11111000011111111110	5.336	3.369	0.107	5.724
6	11111000011111110110	4.971	2.541	0.080	5.741
7	11111000011111110100	5.447	3.481	0.110	5.791
8	11011000001111110100	5.591	3.276	0.104	5.923
9	11111000011111111111	5.245	3.200	0.101	5.673
10	11111000011111110100	5.392	3.341	0.106	5.771

Table 3. Adding 25 fixed strategies to the evaluation procedure, along with the 100 co-evolving GA individuals, causes the GA to produce strategies that can co-operate more with each other, but are not exploited by the more non-cooperative strategies from the enumerative search.

Co-evolution, with the addition of 25 fixed strategies, which are 4 times as likely to be selected into the group of 8 players for 8IPD.

GA strategies play against themselves.

		Mean	Stdev	Stddev of mean	Mean of opponents
0	11111000011111110010	12.575	1.737	0.055	12.740
1	11111000011111110011	12.468	1.939	0.061	12.641
2	10111000011111010010	12.400	2.130	0.067	12.593
3	11111000011111111110	12.557	1.864	0.059	12.709
4	11111000011111110111	12.556	1.488	0.047	12.820
5	11111000011111010110	12.490	1.454	0.046	12.772
6	10111000011111110011	12.392	2.087	0.066	12.568
7	11111001011111111111	13.204	2.457	0.078	10.713
8	11111000011111111111	12.551	1.852	0.059	12.700
9	11111000011111110010	12.560	1.904	0.060	12.718
10	11111000011111110010	12.494	1.835	0.058	12.669

Best 25 strategies from GA search play against a pool of (1) 25 best from enumerative search, and (2) 25 unseen strategies from enumerative search. Note there is little diversity in the GA population.
GA strategies play against pool of 25 seen and 25 unseen strategies from enumerative search.

		Mean	Stdev	Stddev of mean	Mean of opponents
0	11111000011111110010	5.209	3.212	0.102	5.634
1	11111000011111110011	5.494	3.451	0.109	5.828
2	10111000011111010010	5.635	3.120	0.099	6.217
3	11111000011111111110	5.600	3.561	0.113	5.907
4	11111000011111110111	5.187	2.835	0.090	5.966
5	11111000011111010110	5.132	2.762	0.087	5.910
6	10111000011111110011	5.375	3.159	0.100	5.753
7	11111001011111111111	5.447	3.481	0.110	5.788
8	11111000011111111111	5.422	3.340	0.106	5.765
9	11111000011111110010	5.245	3.200	0.101	5.673
10	11111000011111110010	5.392	3.341	0.106	5.771

Table 4. Increasing the importance of the extra 25 fixed strategies causes the co-evolutionary GA to produce strategies that are even more cooperative among themselves, but are still not exploited by the unseen strategies of the enumerative search.

A Systolic Architecture
for High Speed Hypergraph Partitioning
Using a Genetic Algorithm

Heming Chan[1] and Pinaki Mazumder[2]

[1] Intel Corporation, JF1-71, N. E. 25th Ave., Hillsboro, OR 97124 U.S.A.
[2] Dept. of EECS, The University of Michigan, Ann Arbor, MI 48106 U.S.A.

Abstract. We present a systolic array architecture to solve the problem of hypergraph partitioning. The architecture is based on a sophisticated search technique belonging to the class of genetic algorithms. A hypergraph is decomposed into a stream of fine grained, bit string data in which they are propagated into an array of locally connected processing elements. Although each processing element can handle only a few simple bit level Boolean operations, it is shown that the overall connected array forms a powerful hardware partitioning engine in which pipelining and parallelism are fully exploited. Three inner procedures in this GA based solution were parallelized, namely, the fitness evaluation, crossover and mutation operations. A time complexity analysis together with a brief logic block diagrams for the parallel architecture are presented. Simulated results indicated good speedup.

1 Introduction

In this paper, we propose a genetic algorithm for solving the problem of hypergraph partitioning. In addition, a parallel systolic architecture is also proposed to parallelize the inner loops of the algorithm. Given a graph, the problem is to search for a partition of its vertics which optimizes a given cost function. There are numerous applications of this problem: The partitioning of VLSI circuits [1], the subdivision of a computer network into smaller clusters so that the overhead for routing is minimized [13], image segmentation [14], as well as mapping of parallel programs onto parallel machines [5].

The use of genetic algorithms in partitioning is not new. Hulin [7] employed a two step coding scheme for circuit partition and produced excellent results for bit slice circuits. Shahookar [8] applied genetic algorithms to multiway partitioning and obtained better results than the traditional F&M approach [9], and recently, Bui [6] used a fast local optimizer together with a genetic algorithm to solve the problem. Experimental results on graphs from industrial circuit benchmarks are favorable when compared with the recently published result including ratio-cut [1] and spectral methods [10]. Standard genetic algorithms are slow due to the large population size employed. Talbi [5] implemented a genetic based partitioning algorithm in parallel computer and obtained linear speedup. Although many heuristic algorithms had been proposed for hypergraph partitioning [1] [2], little attention has been devoted to the problems from a systolic array implementation point of view. In this paper, a parallel systolic architecture is proposed to parallelize three inner procedures in this GA based solution, namely, the fitness evaluation, crossover and mutation operations. Applications of systolic array originated from the demands of massive computation of real time data required for such areas as communication, signal and image processing [4]. To the best of our knowledge, the mapping of a genetic algorithm to systolic array is new, especially when used as a partitioning engine.

The remainder of the paper is organized as follows. In Sect. 2, we introduce the hypergraph partitioning problem and the essence of a general genetic algorithm. Sect. 3 describes a mapping of the algorithm to a systolic array architecture that parallelizes the fitness evaluation phase. Sect. 4 gives a cellular implementation of the mutation and crossover operators. Experimental studies are presented in Sect. 5 and Sect. 6 concludes the paper.

2   Genetic Algorithms for Partitioning

In this section, the essence of a genetic algorithm will be briefly reviewed. We show how a hypergraph from a partitioning problem can be represented in terms of binary bit strings and hence solved by a genetic algorithm. A hypergraph $H = (V, S)$ consists of a vertex set V and a hyperedge set S which is a set of subset of V. Formally, we define a hypergraph and its partition as follows.

Definition 1. *A hypergraph $H = (V,S)$ consists of a set of m vertices $V = \{v_1, v_2, \dots , v_m\}$ and a set of n hyperedges S, where $S = \{ (v_1, v_2, \dots , v_p) \mid v_1, v_2, \dots , v_p \in V.$ and $v_1, v_2, \dots , v_p$ are connected in H$\}$*

Also, the ordering of the vertices in V is assumed to be fixed throughout this paper. Since a hyperedge is equivalent to a connection in a circuit, we use the term *net* interchangeably for hyperedge in this paper.

Definition 2. *A partition $p = (U, W)$ of a hypergraph $H = (V,S)$ is defined by two sets U and W satisfying $U \cap W = \Phi$, and also $U \cup W = V$. Also we define the set $P = \{ p = (U, W), p$ is partition of H$\}$.*

Problem RP: *Given a hypergraph* $H = (V, S)$ *find a partition* $p = (U, W)$ *s.t. the ratio cut cost R of the cut set C is minimized, where the cut set C is defined as*

$$C(U, W) = \left\{ (v_1, \dots, v_p) \in S \mid \text{at least one of } v_i \in U, v_j \in W \right\},$$

and the ratio cut cost R is defined as $|C(U, W)| / |U| \cdot |W|$

2.1 Genetic Algorithms

Genetic algorithms (GAs), developed by Holland et. al. [3] is an adaptive search strategy based on the mechanics of natural selection in a biological system. It is a highly parallel mathematical algorithm that transforms a set (*population*) of individual mathematical objects (*typically fixed-length character strings patterned after chromosome strings*), each with an associated fitness value, into a new population (*i.e., the next generation*) using operations patterned after the Darwinian principle of reproduction and survival of the fittest and after naturally occurring genetic operations. It has proven to be a robust and effective strategy over a broad range of applications. The basic structure of the algorithm is shown in Fig. 1. It consists of the following main operations:

- A *fitness evaluation* phase that computes the quality of a solution.
- *Crossover*, and *mutation* operations that modify solutions in the current population.
- A *parent selection* operation.
- A *sorting operation* that constructs the next population from the present one.

Table 1 shows the average CPU usage per generation when a sequential machine is used to implement the algorithm. It is clear that the fitness evaluation and the genetic operations dominate the computational time. The parallelization of these two operations is therefore addressed in this paper.

Table 1. Percentage of CPU usage per generation

Fitness evaluation	Genetic operations	Selection	Sorting
64.49%	33.29%	1.4%	0.08%

2.2 Representing the Problem

In order to use a GA, we must represent the problem so that solutions can be specified as binary bit strings. One possible representation for the hypergraph problem is presented in the subsection.

Lemma 1. *Given a hypergraph $H = (V, S)$, $V = \{v_1, v_2, \ldots, v_m\}$, a partition $p = (U, W)$ of H can be uniquely represented by a bit string $(b^1, b^2, \quad b^m) = b$, where for each bit,*

$$b^i = \begin{cases} 0 & ; \text{if } v_i \in U \\ 1 & ; \text{if } v_i \in W \end{cases}$$

Also, we define the solution space B_m as the set of all possible bit strings of length m. We call elements of B_m *solution bit strings*.

The proof of Lemma 1 is straight forward as it is clear that there is a one-to-one mapping between each possible partition and its corresponding bit string representation. This provides a bit string representation of all possible partitions.

Genetic Algorithm ()
Randomly generate a set of p bit strings A_m.
For g iterations **do**
    For number of bit string in A_m **do**
        Select b_1 from A_m randomly.
        Select b_2 from A_m based on its fitness, higher fitness
        has higher possibility to be selected.
        $b_3 \Leftarrow$ apply crossover to b_1, b_2.
        $b_3 \Leftarrow$ apply mutation to b_3 with probability P_m.
        Calculate the fitness of b_3, add it to A_m.
    End for
    $A_m \Leftarrow$ Select the best p bit strings from A_m
End for
End Genetic Algorithm ()

Fig. 1. A Genetic Algorithm

2.3 Details of Genetic Algorithm.

Genetic operators are modifiers of bit strings in successive generations. There are two types of operator, namely the *crossover*, and the *mutation* operator.s. Crossover is a primary operator that combines bit strings in a population to form new bit strings or so called *offspring* . Given two bit strings a and b, a random crossover point is first selected. The bit strings are then divided into two parts, say the left and the right of the crossover point. A new offspring is formed by taking the bits from the left of a together with the right of b. During the process of crossover, bits that carry good partition properties in those strong bit strings will have a higher chance to merge and therefore progressively produce better solutions in each generation.

Mutation provides the algorithm a way to escape from local optimal solutions. During mutation, a bit is randomly selected from the bit string and the value of the bit is changed. The mutation rate, which is the probability of a bit undergoing a mutation operation should be very small, otherwise the algorithm will behave like a random search. Mutation is therefore not a primary genetic operation. Rather it serves to guarantee that the probability of searching a particular subspace in the solution space is never zero. There is fitness associated with each bit string in the population. The fitness of a bit string reflects the quality of its corresponding solution. Since the objective of a genetic algorithm is to maximize the fitness, we define the fitness as the reciprocal of the ratio cut cost R defined in problem RP.

3 Systolic Array for Fitness Evaluation

In this section, a systolic array is presented to parallelize the fitness evaluation phase in the GA. A systolic array organizes itself as a synchronized computational unit through a lattice of identical function modules called processing elements (PE). Unlike general parallel processors, a systolic array is characterized by its regular data flow. Typically, two or more data streams flow through a systolic array in various speeds and directions. Data items from different streams interact with one another at the PEs where they meet. The crux of the systolic approach is that once a data item is brought out from the system memory it can be used effectively at each PE it passes. A higher computation throughput can therefore be achieved as compared to a general purpose processor in which its computation speed may be limited by the IO bandwidth. If a complex algorithm can be decomposed to fine-grained, regular operations, each operation will then be simpler to implement. The simplicity and regularity of the solution lead to a cheaper VLSI implementation as well as higher chip density.

3.1 Cut Operation

To evaluate the fitness of a solution bit string requires counting the number of cuts which is time-consuming and data intensive. In this subsection, we will show how to speed up the cut operation.

Lemma 2. *Given a hypergraph $H = (V, S)$, let $S = \left\{ s_1, s_2, \dots, s_n \right\}$, $V = \left\{ v_1, v_2, \dots, v_m \right\}$, each net $s \in S$, can be uniquely represented by a bit string $e = \left(e^1, e^2, \dots, e^m \right)$, where for each bit,*

$$e^j = \begin{cases} 1, & \text{if } v_j \text{ is connected to net } s. \\ 0, & \text{otherwise.} \end{cases}$$

This bit string is called a *net bit string*. Also, we define N_m be the *input netlist space*, or simply the *input space*, as a set of n net bit strings that are mapped uniquely from the input hyperedge set S. Conceptually, the set represents n corners in a m-dimensional hypercube.

A *one* in the *i-th* bit of a net bit string $e \in N_m$ indicates that the net corresponding to e is connected to the *i-th* vertex. Uniqueness of Lemma 2 can then be easily induced. As in Fig. 2, let $V = \{a, b, c, d, e, f, g\}$ be an ordered set that represents the modules, then $e = (0101001)$ represents a net $s \in S$, connecting to elements b, d and g. Lemma 2 therefore gives a bit string representation of the hyperedges in a graph.

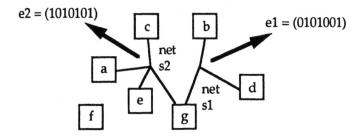

Fig. 2. Bit string representation of netlist.

Definition 3. *The binary cut operator* $\otimes : B_m \times N_m \to \{0,1\}$ *is defined via*

$$b \otimes e = \prod_{i\, st.e^i = 1} b^i$$

where $b = (b^1, \dots, b^m) \in B_m$, *and* $e = (e^1 \dots e^m) \in N_m$. *Here* $\prod$ *is a type of exclusive-or operator, that results in a zero if and only if all the bits under operation are all zero or all one..*

From Lemma 1 and 2, we see that b and e represent a possible partitioning solution and a net in a given hypergraph. The cut operator simply determines if the partition forms a net cut on a net. For example, let $b = (0101001)$, $e_1 = (0101001)$ and $e_2 = (1010101)$, the cut operator applied to b and e_1 will be on the second, fourth, and the seventh bit of b, which is 1,1,1 respectively. The result is, therefore 0. However, $b \otimes e_2$ involves the first, third, fifth and seventh bit of b, which are 0, 0 ,0 and 1 respectively, and since they are not all 1 or all 0, the result is 1. This example is shown in Fig. 3, where b represents a partition of six elements, and e_1 and e_2 represent two nets s_1 and s_2. The partition does cut the net s_2.

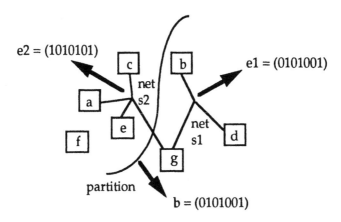

Fig. 3. The cut operation.

Definition 4. *For each $b \in B_m$, we define the size operator $Z(b)$ as the number of ones in the bit string b. For example, if $b = (010010)$, then $Z(b)$ is equal to 2 since there are two ones in b.*

Corollary 1. *Given a hypergraph $H = (V, S)$, it induces two sets B_m and N_m. The partitioning problem is then equivalent to finding a bit string $b \in B_m$, such that the following objective function is minimized.*

$$\frac{\sum_{j=1}^{n} b \otimes e_j}{Z(b) \times (m - Z(b))} \tag{1}$$

where $e_j \in N_m, j = 1,2,..., n$.

Proof. According to Definition 3, $b \otimes e$ gives a 1 if the solution corresponding to a bit string b creates a cut in the corresponding net bit string e. The summation therefore gives the *total* number of net cuts for this solution. It is then obvious to see that Eq. 1 is equivalent to the ratio cut objective. Lemma 1 and 2 establish one-to-one mapping of both the solution space and the input space to their bit string representations. The problems are therefore equivalent.

Corollary 1 provides a way to map fine grained data to a locally connected systolic array, and hence the ratio cut cost can be computed effectively. A simple PE consists of inputs from the left and the top, and outputs from the right and the bottom, as shown in Fig. 4a. In Fig. 4b, m such identical PEs are connected side by side forming a processing array used for the cut operation. The PEs are synchronized in the way each PE takes bit data from inputs, and pass the result to its output at the end of each clock cycle. The input is fed into the array so that the j-th bits of the net bit string e and the solution bit string b are fed into the inputs of the *j-th* PE respectively. The PEs are numbered from left to right starting from 1 to *m*.

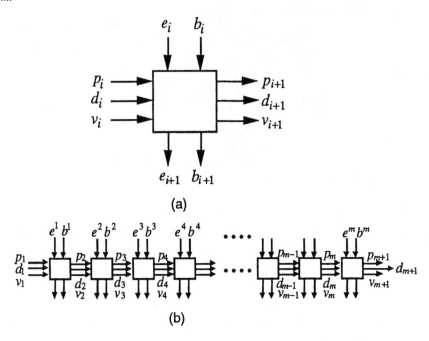

Fig. 4. (a) a processing element. (b) a linear processing elements.

Corollary 2. *Given m linear connected PEs configured as in* Fig. 4b, *with the solution bit string b and the net bit strings e flowing from the top of the PEs, then a cut operation can be performed in m clock cycles.*

Proof. One possible logic implementation of the processing element is,

$$d_{i+1} = d_i + e^i v_i (b^i \oplus p_i) \tag{2a}$$

$$v_{i+1} = v_i + e^i \tag{2b}$$

$$p_{i+1} = b^i \overline{v}_i + p_i v_i \tag{2c}$$

for $i = 1... m$, with $p_1 = v_1 = d_1 = 0$. We would like to show that $d_{m+1} = b \otimes e$. From equation 2b, the signal v of a PE outputs a 1 as long as it gets the first non-zero input net bit. Suppose PE_k is the PE that has the *first* non-zero input net bit, that is $e^j = 0$ for $j = 1$ to $k-1$, and $e^k = 1$. We separate the PEs into three groups, with $j < k$, $j = k$ and $j > k$. For $j < k$, we have, by construction, $e^j = 0$ for $j = 1,...k-1$, and from (2b) and (2c), we have $v_j = d_j = 0$; For $j = k$, we have, by construction, $e^k = 1$, therefore, $v_{k+1} = 1$ and $p_{k+1} = b^k$ Since $v_k = d_k = 0$, then $d_{k+1} = 0$; At last, for, $j > k$, we have $v_j = 1$, and $p_j = b^k$ $j = 1... m+1$, then (2a) becomes

$$d_{j+1} = d_j + e^j(b^j \oplus b^k) \qquad j = k+1, \ldots, m+1 \qquad (2d)$$

which is a recursive definition of the cut operation. Since all PEs are clocked, the operation takes exactly m clock cycles.

3.2   Data Flows in Array

The fitness evaluation in the genetic algorithm requires not only a single cut operation but to find out how many nets are cut by each solution bit string. Furthermore, a large number of solution bit strings in a population need to be evaluated. Let c_i be the total number of cuts for the i th solution bit string in a population, we have to compute a set of p values for each generation, where p is the number of solution in a population.

$$c_i = \sum_{j=1}^{n} b_i \otimes e_j \qquad \text{for } i = 1, 2, \ldots, p \qquad (3)$$

The time complexity of this operation is $O(m \times p \times n)$ per generation.

A systolic array which is used to compute (3) is configured as in Fig. 5. Each row in the array consists of m locally connected PEs, and there are p rows connected vertically forming a $m \times p$ processing array. The p solution bit strings are first loaded into the array. They are stored in such a way that the i-th bit string is stored in the i-th row of the array. The array operates as a synchronized data flow machine in which the net bit data stream is pumped into the array from the top and proceeds down one row at a time in each clock cycle. The appropriate computation is performed when a net bit and a solution bit meet. The net bit strings propagate down in an *angled* manner in which each bit entering a PE is one clock cycle later than its left neighbor. This *angled* configuration allows each processing element to remain active for computation and therefore, its processing power is fully utilized. In addition, there are counters connected at the end of each row collecting cut results for each solution. These counters are called *cut counters*. Fig. 5 shows a snapshot of the propagation of net bit strings through the array. From Corollary 2, it is clear that the i th cut counter in i th row gives the *total* number of cut c_i for solution b_i in $m+i-1$ clock cycles.

Fig. 6 shows three snapshots of the data flowing inside the systolic array. The solution bit strings are first loaded and stored inside the array through the input pins from the top as illustrated in Fig. 6a. The net bit strings are then flow into the array, updating the cut counters after each cut operation (Fig. 6b). The operation is completed when the net bit string completely leave the array. The cut results are collected and are ready to be shifted out from the cut counters (Fig. 6c).

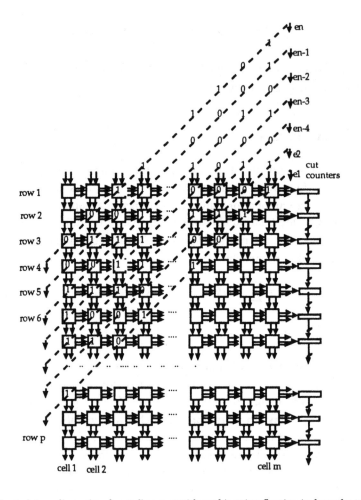

Fig. 5. A two-dimensional systolic array with net bit string flowing in from the top.

Theorem 1. *By using the systolic array, the computation of the total number of cuts for all the solutions in a population takes $m+2p+n$ clock cycles, where p is the size of the population, m is the number of vertices and n is the number of net in a hypergraph.*

Proof, It takes p clock cycles to load and store the solution bit strings into the array. Notice that the cut results for the last generation can be shifted out from below at the same time. From Corollary 2 and Fig. 6c, it is clear that it takes $p+m+n$ clock cycles for computing the cut results. As a whole, it takes $m+2p+n$ clock cycles to complete the operation.

The benefits from the theorem are clear. Since it takes only $m+2p+n$ clock cycles to complete the cut operation, the fitness could thus be computed significantly faster than the $m \times p \times n$ bit operations required in a sequential machine. Moreover, once the inputs have been bought into the array, the processing can be done *without* any data fetching from the system memory, which alleviates the I/O bottleneck.

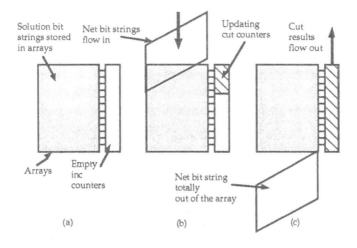

Fig. 6. Signals flow in systolic array.

3.3 Size Operation

The *Size* operation counts the number of ones in a solution bit string, and for each generation, it needs $m \times p$ bit comparisons. To parallelize the operation, we attach a m-bit shift register to each row of the array such that the i-th bit of the register is implemented in the i-th PE. Each solution bit stored in the PE first make a copy of itself to the shift register and the copy is then right shifted at each cycle. A counter is used to collect the number of ones obtained at the right end of the row. It is straightforward to see that once the solution bit string is loaded into the array, it takes m cycles to compute the size of the string. Including loading of the solution data, it takes a total of $m+p$ clock cycles for the entire array. Since this could be done in parallel with the cut operation, the processing time is dominated by the cut which is by far the longest of the two operations, and therefore, the size operation could be completed virtually in no time.

Fig. 7 illustrates the data flowing inside the array. After the solution bit strings are stored in the PEs, the net bit strings enter from the top. Two streams of data flowing from left to right: the data computing the cut and the size operations. They are all collected by counters.

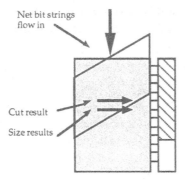

Fig. 7. Signal flow in systolic array, net bit string flowing in, cut and size result flowing out.

4 Cellular Implementations of Genetic Operators.

We propose a fine-grained VLSI implementation for the genetic operators. Since genetic operations are non deterministic, a pseudorandom number generator or PRNG is required. In this section, a cellular automata based PRNG is first introduced. The extensions to crossover and mutation operations are addressed together with simulated results showing the quality of the hardware implementations.

4.1 Random Number Generation

There are many ways of generating good pseudorandom sequences. For fine-grained applications, a cellular automata (CA) based PRNG is normally used because of its simplicity and quality of randomness [11]. A CA-based PRNG is a linear array where the next value at a bit depends only on its present value and the values of the left and right neighbors. Fig. 8 illustrates such a simple one-dimensional CA with the ends cyclically connected. The following update rule is used in our PRNG implementation.

$$a_i(t+1) = a_{i-1}(t) \oplus \big(a_i(t) \cup a_{i+1}(t)\big) \tag{4}$$

This particular PRNG has been extensively investigated by Wolfram [12] and result indicates that each cell output can be considered as an excellent random bit sequence. Also, a comparison study on VLSI implementation area performed in [11] showed that this PRNG gives reasonable small area usage. Our experiments suggested a linear array of size 30 is sufficient for good randomness.

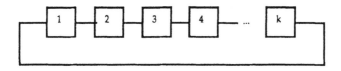

Fig. 8. A simple one-dimensional CA with cyclic boundary condition.

4.2 Crossover Implementation

A block diagram of the crossover implementation is shown in Fig. 9. The crossover registers shown in the diagram are linear, cyclically connected arrays in which the bits from one side of the registers are all zeros (or ones) and the other side are all ones (or zeros). Moreover, the registers are right shifted in each clock cycle. Given two solution bit strings, a crossover could be easily performed by taking the bits from the first (second) bit string if the corresponding cell in an *selected* crossover register is zero (one). The CA-based PRNG described above is used in the implementation. Three bits from the PRNG are used as inputs of a decoder. One of the eight decoder outputs will be active and the corresponding register is selected as the *active* register. Since each bit in the PRNG could be considered as a random sequence, the selection scheme used is also a random process. Moreover, the shifting of the registers warrants crossover points to be uniformly distributed throughout the solution bit string.

The randomness of the crossover point is a good indicator of implementation quality. We simulated the process by using a PRNG of length 30, and eight crossover registers with length 200. The simulated relative frequency distribution shown in Fig. 10 demonstrates a high uniformity throughout the 199 crossover points in a solution bit string.

4.3 Mutation Implementation

Fig. 11 shows a diagram of the mutation operation. The implementation of the mutation operator is very similar to the crossover operator. However, there are two major differences. First, an input *perform* is added to the decoder. The output of the decoder are all zeros if the input *perform* is inactive. If all outputs of the decoder are zero, then no mutation will be performed; Secondly, the eight registers, now called mutation registers, consist of only *one* active bit, since mutation only modifies a single bit in a bit string. The input *perform* is connected to the output of an AND gate having L inputs directly connected to the PRNG cells. It is straightforward to see that the value L controls the rate of the mutation. Since each bit in the PRNG could be considered as a random sequence, the input *perform*, and therefore the output of the decoder, is also a random process. Given a solution bit string, a mutation operation could be easily performed by inverting a bit in the input bit string if the corresponding bit in the *selected* mutation register is active.

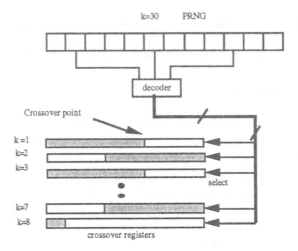

Fig. 9. An implementation of the crossover operator.

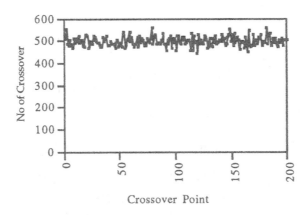

Fig. 10 Simulated crossover shows high crossover uniformity.

We simulated the mutation process by using a PRNG of various length k. Fig. 12 shows how the mutation rate changes with increasing number of AND gate inputs L. Notice that for $k = 30$, the mutation rate for each value of L is closed to the ideal (which is $1/2^l$). The simulated relative frequency distribution shown in Fig. 13 demonstrates a high uniformity throughout the 200 mutation points in a solution bit string.

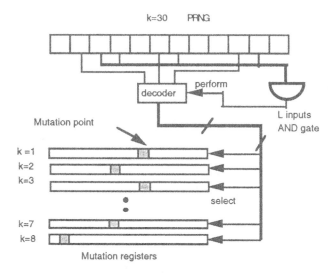

Fig. 11. An implementation of the mutation operator.

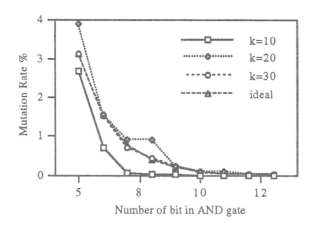

Fig. 12. Mutation rate vs. number of inputs in the AND gate.

5 Performance Analysis

In this section, we study the performance gains when the proposed hardware arrays are employed to accelerate the partitioning process. Since the genetic operations could be done virtually in parallel with the fitness evaluation, we assume that the execution time is dominated by the fitness evaluation. Since we do not have a cellular algorithm for the parent bit string selection as well as the sorting operation to form the next generation, we simply use a software approach instead of a hardware solution for these operations. Also, a clock rate of 10 MHz is used and each register shift takes one clock cycle. Three graphs from industrial circuits were used as testcases. The number of vertices in the circuits are 2379, 1663, 1515, and the number of nets are 3228, 2308, 2189, respectively. We simulated the array using the above assumptions and compared the execution time of the hardware array with a sequential one. Fig. 14 showed the speedup when only the time used by genetic operations together with the fitness evaluation processes are compared. The hardware approach gives up to 250 times speedup. Fig. 15 showed an overall speedup when the entire partitioning process are compared, including the parent selection and population sorting. The overall gain is not very high due to the fact that the population sorting (which run sequentially) becomes increasingly expensive as the population size increases. The parallelization of the sorting and the selection processes are therefore, proposed for future research.

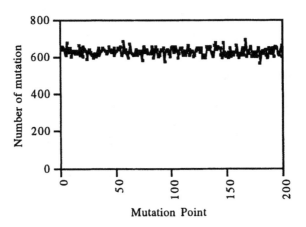

Fig . 13. Simulated mutation shows high mutation uniformity.

6 Conclusion

The progress of the VLSI technology has made the design of complex special-purpose circuits realistic and feasible. In addition, it has encouraged researchers to find hardware solutions to a large class of problems. In this paper, a genetic algorithm is used to solve the hypergraph partitioning problem. A parallel systolic architecture is proposed to parallelize three inner procedures in this GA based solution, namely, the fitness evaluation, crossover and mutation operations. We did a time complexity analysis and also sketched the logic design for the parallel array. Experimental results also indicated order of magnitude speedup. Our results also give new insights in the systolic implementation of other genetic algorithm applications which become increasingly popular. Future research will focus on the parallelization of population sorting as well as the parent selection processes.

Moreover, configurability and fault tolerance issues of this hardware array need further investigation.

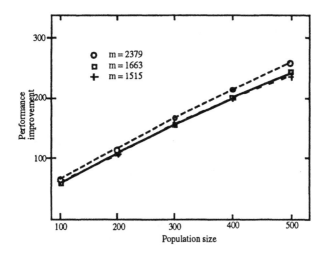

Fig. 14. Speedup for fitness evaluation and genetic operators.

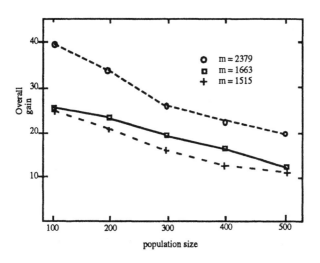

Fig. 15. Overall speedup.

Acknowledgment

The authors would like to thank K. Shahookar for his constructive comments and Intel Corperation for their financial support. They would also like to thank the reviewers for their comments that improves the readability of this paper.

References

[1] Yen Chuen Wei, Chung-Kuen Cheng, Ratio Cut Partitioning for Hierarchical Designs. IEEE Transactions on Computer-Aided Design Vol 10 No 7, July 1991.

[2] Anthony Vannelli, and Scott W. Hadley and Brian L. Mark, An Efficient Eigenvector Approach for Finding Netlist Partitions. IEEE Transactions on CAD. Vol 11 No.7 July 1992.

[3] J. H Holland, Adaptation in Natural and Artificial systems, Ann Arbor, The university of Michigan Press.

[4] J. M. Speiser, H.J. Whitehouse, K. Bromley: Signal Processing Applications for Systolic Arrays. Naval Ocean System Center, San Diego, CA 92152, U.S.A.

[5] E-G Talbi and P. Bessiere, A Parallel Genetic Algorithm for the Graph Partitioning Problem. International Conference on Supercomputing pp. 312-320, 1991.

[6] T. N. Bui and B. R. Moon, A Fast and Stable Hybrid Genetic Algorithm for the Ratio-cut Partitioning Problem on Hypergraphs. 31st, IEEE Design Automation Conference, pp 664-668.

[7] Hulin M, Circuit Partitioning With Genetic Algorithms Using a Coding Scheme to Preserve The Structure of A Circuit. Proceeding of the First Workshop on Parallel Problem Solving from Nature, pp 75-79.

[8] K. Shahookar and P Mazumder, A Genetic Approach to Partitioning, private communication.

[9] C.M. Fiduccia, and R.M. Matteyses, A Linear Time Heuristic for Improving Network Partitions, IEEE Design Automation Conference, pp 175-181, 1982.

[10] L. Hagen and A. Kahng, New spectral Methods for Ratio Cut Partitioning and Clustering, IEEE Transactions on CAD 11 (9), Sept. 1992.

[11] P. D. Hortensius, R. D. McLeod and H. C. Card, Parallel Random Number Generation for VLSI Systems Using Cellular Automata. IEEE Transactions on Computer, Vol 38, No 10 1989.

[12] S. Wolfram, Statistical Mechanics of Cellular Automata. Rev, Modern Phys., vol 55 pp 601-644, 1983.

[13] A. Bouloutas, P.M. Gopal, Some Graph Partitioning Problems and Algorithms Related to Routing in Large Computer Networks. 9th Int, Conf. on Distributed Computing Systems, pp 362-370, 1989.

[14] L. Herault, J. J. Niez, How Neural Networks Can Solve Hard Graph Problems: A Performance Study on the Graph k-partitioning. Neuro-Nimes 89, Int. Workshop on Neural Network and Their Applications. Nimes, France, pp 237-255, Nov. 1989.

Development of Hybrid Optimisation Techniques Based on Genetic Algorithms and Simulated Annealing

Kit Po Wong and Yin Wa Wong

Artificial Intelligence and Power Systems Research Group, Department of Electrical and Electronic Engineering, The University of Western Australia, Nedlands, Western Australia 6009

Abstract. This paper develops a hybrid optimisation algorithm, GAA, by combining the genetic algorithms (GAs) approach and the simulated-annealing technique (SA). The combination facilitates the introduction of more diversity into the population and prevents the problem of premature convergence. To counter the adverse effects of mutation, two effective measures are developed and included in the combined GA/SA method. This algorithm is then further developed to minimise the memory requirement. The revised hybrid algorithm GAA2 is analysed and compared to GAs, GAA and SA. A guideline for setting the parameters for executing GAA2 is also established. The performance of the developed algorithms are demonstrated through their applications to the hydro-thermal scheduling problem in power systems.

1 INTRODUCTION

Genetic algorithms (GAs) [1,2] are an adaptive searching technique for solving optimisation problems based on the mechanics of natural genetics and natural selection. Owing to their ability to seek the global or near-global optimum solution, work has recently been performed to apply GAs to numerous areas in engineering including image processing [3,4], VLSI design [5-7], robotics systems [8], transportation [9, 10], and power engineering [11-15].

The success of the application of GAs to an optimisation problem depends on the design of (a) representation of chromosomes, (b) fitness function, (c) method of crossover and (d) mutation operation. In the representation of chromosomes, appropriate coding methods for chromosomes are essential. Presently the binary coding method is most widely used and it is suitable for discrete optimisation problems. However, for problems dealing with parameters having continuous values, this coding method is inappropriate due to its inherent discretisation problem [13, 16].

The success of GAs also depends on the diverse information held in the chromosomes. When the diversity is lost well before the global optimum solution is found, the performance of GAs deteriorates and their solution processes converge prematurely [17-19]. Although some methods for preventing premature convergence have been suggested by Booker [20] and by other researchers, more efficient techniques are worthwhile to be developed.

Moreover, the performance of GAs is subject to the mutation operation to some extent. While the mutation operation adds new information to a chromosome, it can also destroy useful information originally held in the chromosome [21]. The performance of GAs can be improved if a mechanism can be incorporated in GAs so that the positive effects of mutation will be retained but the adverse effects will be eliminated.

To improve the applicability and performance of GAs when they are applied to optimisation problems with parameters which can take continuous values, this paper first reports a floating-point number coding method for the representation of the chromosomes to overcome the discretisation problem. The paper then develops two hybrid optimisation algorithms using GAs and simulated-annealing (SA) [22, 23]. SA is an optimisation technique which simulates the physical annealing process of a molten particle. In the hybrid algorithms, the SA techniques are utilised and incorporated into GAs for the alleviation of the premature convergence problem and for the reduction of the adverse effects of the mutation operation. Based on the developed hybrid algorithm GAA, and by restricting the population size to 2, another hybrid algorithm GAA2, with minimum memory requirement is also developed. The hybrid algorithms are analysed, and their performance are demonstrated through an example in which the optimum generation schedule of a mixed hydro and thermal power system is sought.

2 GENETIC ALGORITHMS

GAs are search algorithms in which the search is conducted using information of a *population* of candidate solutions so that the chance of the search being settled in a local optimum can be significantly reduced. Four essential components need to be designed in applying a GA for an optimisation problem.

1. Candidate solutions representation
   In GA, a candidate solution, which is referred as a *chromosome*, is represented by a string of finite length. Although the values of the elements in the string can be represented by special symbols, [24], they are usually represented by a binary bit string [1, 2, 11, 13, 14]. While representation by special symbols is restricted in general application, the binary bit string representation fails to reflect the continuous nature of the value of the elements, as reported in References 12, 13 and 16.

2. Crossover operator
   The function of the crossover operator is to generate new or child chromosomes from two parent chromosomes by combining information extracted from the parents. The single-point crossover [1, 2] is usually employed in the crossover operation in GA. In this method, for a chromosome of length l, a random number m between 1 and l is first generated. Two child chromosomes are formed by swapping the first m elements of the parents with one another.

When a random number generated between 0 and 1 is less than the preset value of the probability of crossover, crossover will be applied to two parent chromosomes. It can be anticipated that a high crossover rate allows more thorough exploration of the solution space, and reduces the chances of settling for a local optimum. However, if the rate is too high, it will result in the wastage of computation effort in exploring unpromising regions of the solution space. A commonly adopted crossover probability is 0.6 for a population size of 100 [21].

3. Mutation operator

When mutation is applied to a chromosome, new information is introduced into the chromosomes by changing the value of one of its elements. To decide which element is to be changed, a way adopted widely is to generate a random number m between 1 and l and then make a random change in the mth element of the string. When a random number generated between 0 and 1 is less than the preset value of the probability of mutation, a chromosome will be mutated.

The frequency of the mutation operation will affect the performance of GA. If the frequency is too high, excessive new information is introduced and valuable 'old' information may be destroyed in the chromosomes. On the other hand, if the frequency is low, the chromosomes remain virtually unchanged and little information will be added to assist the search in GA. Typically, the probability of mutation is 0.001 for a population size of 100 [21].

Although the mutation operator can, to some extent, introduce some new information into a chromosome and increase the diversity of the chromosomes, it cannot completely eliminate the problem of premature convergence which has been described in Section 1. Moreover, the mutation operation can destroy useful information in chromosomes. Methods for reducing the adverse effects of mutation will be developed in Section 5.2.

4. Fitness function

The fitness function evaluates the 'fitness' of a chromosome. The function can be defined in terms of the objective function of the optimisation problem. The search for the global optimum solution is then equivalent to finding the chromosomes having the maximum fitness. Amongst other factors, the accuracy of the optimum solution found by the GA approach is dependent on the design of the objective function and the fitness function.

2.1 Basic Genetic Algorithm

The solution procedure of a basic GA (BGA) can be summarised below.

1. The initial population of chromosomes is either specified or generated randomly. This set of chromosomes is referred to as the *current* generation. A commonly adopted population size, s, is 100 [21].

2. The fitness of each chromosome is evaluated by the fitness function. If a chromosome has achieved the desired fitness, it is considered to be the optimum solution and the solution process is terminated. If the desired fitness is not known, the process can be halted when a homogeneous population is achieved. This terminating criterion, however, imposes the overhead in checking for identical chromosomes at the end of each generation. The process can also be stopped when the maximum allowable number of generations has been executed.

3. Generate the next generation of s chromosomes in the following way:

    Step 3.1 For each chromosome, calculate its fitness measure over the entire population of the current generation. Assign the fitness measure of the chromosomes as the probabilities for the chromosomes to be selected for reproduction.

    Step 3.2 Reproduction: select from the current generation in a stochastic manner two chromosomes as parents.

    Step 3.3 Crossover: with the specified crossover probability, Prc, apply crossover to the two selected parents in the current generation when the value of a random number generated between 0 and 1 is less than Prc. Otherwise, the two parents are retained and are taken as the child chromosomes in the next generation.

    Repeat the selection step in Step 3.2 and the present step until s child chromosomes are formed for the next generation.

    Step 3.4 Mutation: for each chromosome in the next generation, apply mutation to the elements of the chromosome when the mutation probability, Prm, is greater than the value of a randomly generated number between 0 and 1. Otherwise, the chromosome will remain intact.

4. The next generation formed in Step 3 is now taken to be the current generation. New generations are produced by repeating the solution process starting from Step 2.

In the above procedure, Step 3 is responsible for generating a generation of s chromosomes and it can be regarded as one iteration in the solution process. The 'number of generations' in Step 2 can be conveniently expressed as 'number of iterations'. The flowchart of BGA is shown in Fig. 1. In the figure, k is the iteration counter; p is the counter for the number of chromosomes generated and rand [0,1] denotes a random number between 0 and 1.

2.2 Incremental Genetic Algorithm

A variant of BGA which can find the global optimum solution earlier in the solution process than BGA is the Incremental Genetic Algorithm (IGA) [25-28]. The major difference between IGA and BGA is in Step 3.3. In the IGA approach, the two child chromosomes produced by the selected parents indicated in block C in the flowchart of Fig. 1 are included in the current generation instead of in the next generation indicated in block D of the flowchart. Therefore, there is no distinction between the next generation and the current generation. The above change can be reflected in the flowchart of Fig. 1 by removing the flow path from block A to block B and inserting a new path from block A to block B'. After s chromosomes have been generated, the mutation operator is then applied to them as in Step 3.4 to complete an iteration.

Each child chromosome generated will replace a chromosome, which has a smaller fitness measure and is selected from the existing population using the roulette wheel technique [21]. The two child chromosomes will compete with the remaining chromosomes in the current generation to be selected as parents after the normalised fitness of all chromosomes are evaluated.

3 ELIMINATION OF DISCRETISATION ERROR

Chromosomes are usually represented by strings of binary bits [1, 2, 11, 13, 14] in GAs. When the binary coding method is adopted for coding continuous parameters, discretisation errors will occur and inaccurate solutions will result. Although the discretisation problem has been addressed in Reference 13, no method has been proposed in that reference to alleviate the problem. The authors [29] have proposed a floating-point number coding method for GAs for the elimination of the discretisation problem. In the proposed method, a chromosome is coded using a string of floating-point numbers as shown in Fig. 2. Each floating-point number is an element in the chromosome and the length of the chromosome is given by the number of elements. By this representation method, the need to convert binary numbers into their decimal equivalents or vice versa is avoided. The effects of reproduction, crossover and mutation, on the growth rate of schemata when the chromosomes are coded using floating-point number, have been investigated by the authors [29]. It has been found that under the new coding method, fitter chromosomes propagate and grow at a higher rate in successive generations under the binary coding method.

4 COMBINING GA AND SA

The performance of BGA and IGA relies on the diverse information held in the chromosomes. When diversity is lost well before the global optimum solution is found, the performance of these approaches deteriorates and their solution processes converge prematurely. The phenomenon of premature convergence arises owing to the fact that the probability of selection of parent chromosomes cannot be strictly observed for a finite size of population [20]. This leads to some selection errors in selecting parents. This problem is more prominent with IGA than with BGA as the selection error is propagated immediately into the generation process for new chromosomes.

Although the mutation operator can, in a small extent, introduce some new information into a chromosome and increase the diversity of the chromosomes, it cannot eliminate the problem of premature convergence completely. Mutation can also destroy useful information in the chromosomes in spite of its positive effects [21]. To alleviate the problem of premature convergence, some methods have been proposed by other researchers and they are discussed in Section 5.1 where a new method will be developed.

One of the possible ways to improve the performance of the GAs is to combine GAs with other techniques. In the recent method proposed by Lin, Kao and Hsu [30], SA techniques are combined with a GA in the following way. Starting from a population of chromosome formed by the GA operators, the SA solution generation mechanism and solution acceptance tests are applied to each chromosome to form a new chromosome. The new chromosomes are regarded as chromosomes in the next generation. This process is then repeated. It is terminated when a criterion designed by Lin et al is met. The main disadvantage of this approach is that information held in the chromosomes of the current generation may be destroyed by the solution generation mechanism of SA and therefore may not be inherited by the new chromosomes in the next generation.

In Esbenen and Mazunder's [31] attempt to combine GA and SA, the mutation operation is controlled by the acceptance test of SA and the pure GA is first executed until it stagnates. The GA is then executed with increased mutation rate and decreased population size. This process is repeated until there is only one chromosome. At this point the algorithm is switched to the SA. However, in this method, the adverse effects of mutation cannot be eliminated completely because the magnitude of change to a chromosome due to mutation is not regulated.

From the above discussions, a systematic method for introducing diversity into the chromosomes and a better method to form hybrid GA/SA algorithms need to be developed. This paper proposes a new way to enhance GA using SA. In Section 5, an approach, GAA, based on IGA and SA will be developed. The property of IGA that it can find the global optimum solution early in the solution process is retained in GAA. In the algorithm, premature convergence is eliminated and the destructive effects of the mutation operator are reduced using the perturbation technique and the probabilistic method for accepting new candidate solutions in SA. The SA method is first outlined in the following section.

4.1 Simulated Annealing Method

The optimisation process in SA is essentially a simulation of the annealing process of a molten particle. Starting from a high temperature, a molten particle is cooled slowly. As the temperature reduces, the energy level of the particle also reduces. When the temperature is sufficiently low, the molten particle becomes solidified. Analogous to the temperature level in the physical annealing process is the iteration number in SA. In each iteration, a candidate solution is generated. If this solution is a better one, it will be accepted and used to generate yet another candidate solution. If it is a deteriorated solution, the solution will be accepted when its probability of acceptance is greater than a randomly generated number between 0 and 1. The probability of acceptance, $\Pr(\Delta)$, can be calculated from [32]

$$Pr(\Delta) = [1 /((1 + \exp (\Delta / T))] \tag{1}$$

where Δ is the amount of deterioration between the new and the old solutions and T is the temperature level at which the new solution is generated.

In forming the new solution, the old solution is perturbed according to a probabilistic distribution function. In the application example in Section 8, the Gaussian probabilistic distribution function (g.p.d.f) is employed. The mean of the g.p.d.f. is taken to be the old solution and its standard deviation is given by the product of the temperature and a scaling factor γ. The new solution is formed by adding the amount of perturbation to the old solution. The amount of perturbation is dependent on the temperature when the scaling factor γ is kept at a constant value.

In an iteration, the procedure for generating and testing the candidate solution is repeated for a specified number of trials. The last accepted candidate solution is then taken as the starting solution for the generation of candidate solutions in the next iteration. In the next iteration, the temperature is reduced according to [22]

$$Tk = r(k-1) \, To \tag{2}$$

where To and Tk are the initial temperature and the temperature at the kth iteration respectively and r is the temperature reduction factor. The solution process continues until the maximum number of iterations is reached or the optimum solution is found.

A flowchart illustrating SA is shown in Fig. 3. In the figure, k is the iteration counter and t is the trial counter. A SA-based economic power dispatch algorithm has previously been developed by Wong and Fung [33].

5   THE GAA HYBRID ALGORITHM

By considering that the number of child chromosomes s generated in one iteration in IGA and the maximum number of trials in an iteration in SA are equivalent, the IGA and SA can be combined to form a new algorithm GAA. In GAA, s children are generated in an iteration by the method described in IGA and are considered to be generated at the same temperature level. The flowchart of GAA is shown in Fig. 4.

5.1   Preventing Premature Convergence

To prevent premature convergence, Booker [20] has proposed the 2-point crossover method to solve the problem. However, the amount of diversity introduced into the chromosomes was not regulated during the solution process. Wilson [34] has proposed another method. In his method, the crossover rate is varied according to the value of an entropy measure over the population. When the diversity is low, the entropy measure will be low and the crossover rate will be increased. This method may not be effective because it may be too late to increase the crossover rate especially when the diversity of a population is already low.

Maudlin has also proposed the uniqueness approach [35]. In this approach, the mutation rate is monitored and increased when the diversity of chromosomes begin to diminish. In addition, the amount of increase of the mutation rate is reduced as the solution process progresses. The disadvantages of this approach are that it requires more computing time because of the required evaluations of the degrees of homogeneity of the chromosomes and it can hinder useful information of the parents being inherited by the children. Moreover, a systematic method for changing the mutation rate has not been proposed by Maudlin.

In the present work, to prevent premature convergence in the solution process of GAA, before a selected chromosome is replaced by a child chromosome as indicated in blocks A and B in the flowchart of Fig. 4, a set of replacement criteria is first checked.

1. if the selected chromosome is the fittest generated so far, retain it, otherwise,
2. if the child chromosome is fitter than the selected chromosome, replace the selected by the child, otherwise,
3. if the probability of replacement of the selected chromosome is greater than a random number generated between 0 and 1, replace the selected by the child, else retain the selected.

If all the replacement criteria are not satisfied, another chromosome will be selected and tested for replacement. In criterion (3), the probability of replacement of the selected chromosome is calculated using equation (1). In the present context, Δ in equation (1) is the amount of deteriorated fitness measure of the child chromosome as compared to the selected chromosome for replacement. The probability of replacement will be low when the temperature level is low.

By criterion (3), some fitter chromosomes will be replaced by the child chromosomes during the entire solution process. But the chance for the fitter chromosomes to be replaced is greatly reduced towards the end of the solution process when the temperature level is low. In this way, sufficient diversity of chromosomes can be maintained and premature convergence can be eliminated.

5.2 Countering Adverse Effects Of Mutation

To preserve the positive effects and to counter the adverse effects of the mutation operator, two mechanisms are included in GAA. The first mechanism regulates the degree of change to a chromosome when it is mutated. In this mechanism, after an element in the chromosome is chosen, the value of the element is changed by perturbing its present value according to the Gaussian probability distribution function. This is indicated in block C in the flowchart of Fig. 4. Since the amount of perturbation to the present value decreases as the iteration increases, that is, when the temperature level decreases, the change in value of the selected mutated element is also reduced.

The second mechanism regulates the frequency of the mutation operation using the replacement criteria in the solution process. This is indicated in block D of Fig. 4. In this mechanism, a chromosome is replaced by its mutated chromosome when replacement criteria (2) or (3) is satisfied. It will be retained if it is not replaced. It is also retained if it is the fittest chromosome generated, as stated in replacement criterion (1). As the solution process progresses, the chance for a chromosome being replaced by its mutated chromosome with deteriorated fitness decreases.

6   THE GAA2 HYBRID ALGORITHM

Like BGA and IGA, the GAA approach possesses s chromosomes in a population. Typically, the population size s is set to 100 [21] and the computer memory requirement is high. This requirement can be reduced to a minimum by adopting a population size of 2 throughout the entire solution process. The GAA approach modified in this way is hereafter referred to as the GAA2 approach.

The solution procedure of GAA2 is described by the flowchart in Fig. 5. The procedure starts with an initial temperature level and two parent chromosomes. There are only two chromosomes in the population and they are always taken as the parents for producing two child chromosomes as indicated in block A of the flowchart. The replacement of the parents by the children indicated in block A ensures that there are only two chromosomes in the population. Also in block A, the fittest chromosome so far generated is stored outside of the population. The mutation operation indicated by block B in Fig. 5 will be initiated when a crossover has taken place. The temperature level will be reduced when g child chromosomes have been formed. The parameter g is here termed as the *pseudo-population* size. In order to preserve the elite chromosome, that is the fittest chromosome ever generated, it is re-introduced into the population using the procedure in block C before the temperature level is reduced in readiness for the start of the next iteration in the solution process.

The procedure for generating 2 children and for the replacement of the parents by their children in block A is described in the flowchart in Fig. 6. Before any parent chromosome is replaced, the child chromosome is checked to ensure that it is distinct from the parents. In the case that a child is identical to a parent, it is discarded. The checking and the subsequent actions are detailed in Fig. 6. In the figure, when comparing the chromosomes, the symbol '=' takes the meaning of 'is identical to'.

The replacement of a parent by a child is performed according to replacement criteria (2) and (3). Replacement criterion (1) is not adopted so that even if one of the two parents is the fittest chromosome generated so far, it will also be considered for replacement. This allows more diversity to be maintained in a small population of size two. However, the fittest parent is re-introduced to the population by the mechanism in block C of Fig. 5.

In block B of Fig. 5, the procedure for mutating a chromosome and the checking for its replacement by the mutated chromosome is similar to that of the GAA approach. However, replacement is inhibited if the chromosome and its mutated chromosome are identical. Also if the mutated chromosome is the fittest chromosome generated so far, it will replace the fittest chromosome stored outside of the population. The procedure is described in the flowchart of Fig. 7.

The procedure in block C of Fig. 5 for re-introducing into the population the fittest chromosome which is held outside of the population is described in the flowchart of Fig. 8. A randomly generated number which can take the value of 0 or 1 is used to decide whether the fittest chromosome need to be re-introduced. The re-introduction procedure is bypassed if it is decided not to do so, otherwise, a chromosome is randomly selected from the two chromosomes in the population. Provided that the unselected chromosome is not identical to the fittest chromosome stored, the selected chromosome is replaced by the fittest chromosome.

In comparison to GAA, GAA2 does not require any procedure for selecting parents and it has a very small number of crossover operations since the population size is kept at its minimum. The assurance of sufficient diversity in the chromosomes and the introduction of new information to them through mutation can be achieved by increasing the value of the probability of mutation by a positive factor. In the study example in Section 8, this factor is set to 10. Although a higher mutation probability leads to more mutation operations, GAA2 can still be the fastest approach amongst all the approaches developed.

7   COMPARISON OF THE ALGORITHMS

7.1   Comparing GAs, GAA And GAA2 with SA

One of the major differences between GA-based algorithms and SA lies in the method of generating new candidate solutions. In SA, they are formed by perturbation according to a probabilistic distribution function and in GA-based algorithms, they are produced mainly by crossover. The validity of the candidate solutions depends on whether they can satisfy the equality and inequality constraints in an optimisation problem.

An approach has been proposed by Wong and Fung [33] and by Wong and Wong [12] to deal with the equality constraints. The approach is described briefly here by considering the case that there is only one equality constraint in the problem. One of the parameters in the problem is chosen randomly as a *dependent* parameter and its value is found after values of the remaining parameters are set or found. The values of the *non-dependent* parameters can be found by the perturbation technique of SA or the crossover operation in GAs.

When the values of the parameters in the chromosomes are feasible and satisfy the inequality constraints, by crossover, parameters in any child chromosome produced will have feasible values. This means that the candidate solution is always feasible unless the value of the dependent parameter is found to be out of its limit according to the inequality constraints. In contrast to this, there is no guarantee that the candidate solutions found by perturbation in SA will be feasible.

From the above discussions, it is clear that GAs, GAA and GAA2 have larger capability than SA to deal with problems which have large sets of inequality constraints. Their computational requirements are much lower than that of SA since very small amount of their computational effort will be spent on generating infeasible solutions.

7.2 Comparison Of Performance Between GAA2 and GAs

Since the hybrid algorithm GAA2 holds two chromosomes in a population and they are always used for reproduction and crossover, GAA2 therefore requires the least memory amongst all the GA-based algorithms and it is computationally faster. The following comparison of performance between GAA2 and the other algorithms is concentrated on the decay rates of the schemata and the diversity of the chromosomes. A guideline for setting the parameter values for the execution of GAA2 is also developed in Section 7.2.2.

7.2.1 Comparison of decay rates

The principal difference between GAA2 and other GAs is that in GAA2, a weaker new candidate chromosome replaces a fitter chromosome provided that it passes the probabilistic replacement test as described in Sections 5.1 and 5.2. The inclusion of the replacement test in GAA2 will affect the decay rate of chromosomes whose fitness measures are below average. In terms of the concept of schema [2], it is shown below that weaker schema in GAA2 will decay at a faster rate than GAs.

For GAs, it has been established in Reference 2 that the number of chromosomes, which have less than average fitness and belong to schema S, will decay from generation g to the next generation (g+1) due to the effects of reproduction, crossover and mutation according to

$$F(S, g+1) = [F(S,g) * f(S) / fa] * (1 - Prc * Pd) * (1 - d * Prm) \qquad (3)$$

where $f(S)$ is the average fitness of schema S; $F(S,g)$ is the number of chromosomes in generation g belonging to S; fa is the average fitness of the entire population of chromosomes in generation g; Prc is the probability of crossover; Pd is the probability of schema S being destroyed in crossover; Prm is the probability of mutation; where d is the order of schema S. The three bracketed terms on the right hand side of the above equation describe the effects of reproduction, crossover and mutation respectively. The combined effect is that the schema with less than average fitness and long defining length will decay exponentially [2].

In GAA2, as there are only two chromosomes in the population and they are always selected for reproduction, the first bracketed term in equation (3) becomes F(S,g). However, after crossover and also after mutation, new chromosomes will replace the parent chromosomes with a probability of replacement.

Let the probability of replacement after crossover be Rc and that after mutation be Rm. The second and the third bracketed terms of equation (3) become
(1 - Prc * Pd) *Rc and (1- d * Prm) *Rm respectively. Since Prc in GAA2 is
always equal to 1, the combined effect of crossover, mutation and replacement
is described by

$$F(S, g+1) = [F(S,g)] *(1 - Pd) * (1- d * Prm) *Rc *Rm \qquad (4)$$

From equations (3) and (4), provided that (Rc * Rm) is increasingly less than
(f(S)/ fa), the schema with less than average fitness in GAA2 will decay at a
faster rate than that in GAs. From equation (1), however, Rc and Rm decay
exponentially as the solution process progresses. Even if (f(S)/ fa) decays at the
same rate with Rc or Rm, it will decay at a lower rate than (Rc * Rm). It follows
that generally less fit schemata will decay at a faster rate in GAA2 than in GA.

7.2.2 Comparison of diversity in chromosomes While almost all aspects in
GAA2 and GAA are identical, the principal difference lies in the allowed number
of chromosomes in the population and in the case of GAA2, the number is fixed
at 2. It can be anticipated that if all the parameter settings for the execution
of the two algorithms are the same, then GAA will have greater diversity in
chromosomes than GAA2. However, with proper parameter setting, GAA2 will
have a comparable, or even greater diversity than GAA. Since the parameter
settings suitable for the execution of GAs can be adopted for the execution of
GAA, the above comments on the difference in diversity of chromosomes between
GAA and GAA2 are also applicable to that between GAs and GAA2. Starting
with the set of appropriate parameter settings for the execution of GAs or GAA,
an expression serving as a criterion for the setting of the parameters for GAA2
so that equal or higher diversity of chromosomes can be achieved is developed
below.

In the initial population, assuming that values of all the elements in the
chromosomes are distinct, the measure of the diversity of chromosomes, Iso, in
a GA can be given by m*ps, where m is the number of element in a chromosome
and ps is the population size. Similarly, in GAA2, the measure I2o is 2*m.

When the dependent element approach for satisfying the equality constraints
in Section 7.1 is adopted in GAs or GAA, new values of the dependent elements
are evaluated after the crossover operations. Owing to this, diversity is added to
the chromosomes. The measure of this diversity, Cg , from the 1st generation to
the gth generation of the optimisation process can be given by

$$Cg = Prg * ps * Prc \qquad (5)$$

where Prg is the summation of the probability of the chromosomes being completely distinct from the 1st to gth generation. The probability in any generation
i is 1 when all chromosomes are completely distinct and is zero when they are
identical. Prc is the probability of crossover.

In the case of GAA2, for each pseudo-population, only distinct child chromosomes formed by crossover, and with their dependent element values evaluated, will be accepted after passing the replacement test. Therefore for a pseudo-population size of ps2 and a unity probability of crossover, the measure of diversity due to new values of the dependent elements, C2 , is

$$C2 = ps2 * g2 \tag{6}$$

where g2 is the total number of generation in GAA2 and ps2 is the pseudo-population size.

Similarly, the measures of diversity due to new values introduced by the mutation operation in GAs and GAA2, Mg and M2 , are respectively given by

$$Mg = Prg * ps * Prm \tag{7}$$

$$M2 = ps2 * g2 * Prm2 \tag{8}$$

in which Prm and Prm2 are the probabilities of mutation for GA and GAA2 respectively.

Taking into account the diversity in the initial population and that due to the values of the dependent elements after crossover and mutation, the combined measure of diversity in a GA process, Db , can be formed from

$$Db = m*ps + Cg + Mg \tag{9}$$

The corresponding combined measure in a GAA2 process, D2 , is

$$D2 = 2*m + C2 + M2 \tag{10}$$

For the measure of diversity of GAA2 to be equal to or greater than that of GA, D2 >= Db. From this and from equations (5) to (10), an expression relating the parameter settings for GAA2 to the settings for GA can be derived and is

$$g2 * ps2 * [1 + Prm2] >= m * [ps - 2] + Prg * ps * [Prc + Prm] \tag{11}$$

Taking the extreme case that the chromosomes in all generations are all distinct, Prg of GA will take the value of the total generation G, and the above inequality will become

$$g2 * ps2 * [1 + Prm2] >= m * [ps - 2] + G * ps * [Prc + Prm] \tag{12}$$

When the length of the chromosome, m, is known and the values of the parameters ps, G, Prc and Prm given for the execution of GA, the values of g2, ps2 and Prm2 for GAA2 should be set such that expression (12) is satisfied. Since in general, Prm2 << 1, the above expression can be approximated by

$$g2 * ps2 >= m * [\, ps - 2\,] + G * ps * [\, Prc + Prm] \qquad (13)$$

To enhance higher diversity, Prm2 in GAA2 can be set to (k * Prm) where k is a positive number, the value of which is greater than one.

8 APPLICATION EXAMPLE

The GA algorithms and the developed hybrid algorithms have been implemented using the C programming language and the software systems are run on a PC/486 computer with an i860 co- processor. They are applied to solve the hydro-thermal scheduling problem in a mixed hydro/thermal power system [36]. The primary objective of short-term hydrothermal scheduling is to determine the amount of hydro and thermal generations to meet the electrical load demands in a schedule horizon so that the fuel cost required to run the thermal generators can be minimised. In the application example, the schedule horizon is 3 days and it is divided into six 12-hour intervals. The thermal system is represented by an equivalent generator unit.

A candidate schedule of the thermal unit in the six intervals is represented by a chromosome. The chromosome therefore has a length of 6 and the value of m in expression (13) is also 6. Each of the six elements in the chromosomes is the power production level of the thermal unit. The fuel cost function, in unit of dollars per hour, of the equivalent thermal unit at the power production level Gs is :

$$f(Gs) = 0.00184 \; Gs^2 + 9.2 \; Gs + 575 \qquad (14)$$

The lower and upper operation limits of the unit are 150 MW and 1500 MW respectively. Assuming that the electric loss from the hydro plant to the load is negligible, the specified load demand at any interval must then be met by the sum of Gs and Gh, where Gh is the amount of hydro power generation. To produce the hydro power generation level Gh, the required discharge rate q in acre-ft per hour of the water flowing from the reservoir to the hydro plant is given in equations (15) and (16) below. Equation (15) is for Gh in the range from 0 to 1000 MW and equation (16) is for the range from 1000 MW to 1100 MW.

$$q = 4.97 \; Gh + 330 \qquad (15)$$

$$q = 0.05 \; (Gh - 1000)^2 + 12 \; (Gh - 1000) + 5300 \qquad (16)$$

The volume of water in the reservoir at the jth interval is given by

$$Vj = Vj\text{-}1 + (rj - qj - sj)nj \qquad (17)$$

where rj, qj and sj are respectively the water in-flow rate; water discharge rate and the spillage rate in interval j the duration of which is denoted by nj. For the present problem, the initial and final volumes of water in the reservoir are given and are 100,000 acre-ft and 60,000 acre-ft respectively. The minimum and maximum volumes of water in the reservoir are restricted to 60,000 acre-ft and 120,000 acre-ft respectively in all intervals. The water inflow rate is assumed to be constant and is 2000 acre- ft/hr and the spillage rate is discounted.

Owing to the randomness in GA, GAA and GAA2, the algorithms are executed 30 times when applied to the test system. For the present application study, the adopted values of the parameters for the execution of the algorithms are summarised in Table 1. The parameter values for the BGA, IGA and GAA algorithms in Table 1 are the same as those suggested in Reference 18, except the number of iterations. The number of iterations are found experimentally.

The settings of the number of iterations, g2, the probability of mutation, Prm2 and the pseudo population size, ps2, for GAA2 in Table 1 are determined experimentally and are found to observe the inequality in expression (13). When BGA, IGA and GAA are applied to the present example, their population sizes are reduced from 100 down to 20 but their performance in seeking for the global optimum schedule are inferior to that of GAA2. Comparison of study results in Section 8.1 are therefore based on the population sizes tabulated in Table 1.

Table 1. Parameter values for the execution of the new algorithms

algorithm	BGA, IGA, GAA	GAA2
population size (ps)	100	2
no. of iterations (G, g2)	100	300
probability of crossover (Prc)	0.6	1
probability of mutation (Prm, Prm2)	0.001	0.01
pseudo population size (ps2)	–	40

From the results of the experimental tests conducted by the authors, it is observed that, the temperature must be reduced slowly, that is the temperature reduction rate r should be close to unity, so that the solution process will not settle in a local optimum. The temperature reduction rate r in GAA, GAA2 and SA are all set to a value of 0.98.

For GAA, the initial temperature and the scaling factor γ are set respectively to the values of 50000 and 0.01. The corresponding values used in GAA2 are 25120 and 0.02. For the present study, the parameter settings for SA are : The initial temperature is 50000 and the scaling factor γ is 0.004. The number of trials in a SA iteration is 10000 and the number of iterations is 200. These parameter settings are all established by experimentation.

8.1 Study Results

The best short term hydro-thermal schedules found by the algorithms and the solutions found by the steepest-descent gradient method [36] are tabulated in Table 2. In the table, '1st int.' is the 1st interval starting at 12.00 am and ending at 12.00 pm. '2nd int' is the 2nd interval, which starts at 12.00 pm and ends at 12.00 am. The best schedule solution determined by the SA-based scheduling algorithm recently developed by the authors [37] is also summarised in the table.

The results in Table 2 show that the generator loadings in the optimum schedule found by the gradient approach lead to a higher fuel cost than those found by the other algorithms. Among them, IGA and GAA2 provides the best schedule result with a cost of 709862.055 and 709862.086 respectively.

The worst fuel costs found by the algorithms in 30 executions are summarised in Table 3. The increment of these costs with respect to the cost determined by the gradient approach are given in percentage in the table. The results show that GAA2 and GAA can determine the best solutions and their worst fuel costs are better than that of the gradient approach, BGA, IGA and SA. The worst cost solutions determined by BGA and IGA are only marginally higher than the gradient approach. Among the algorithms, GAA2 is the most promising and SA is the less reliable.

The execution times in 30 executions of the algorithms are summarised in Table 4. The average execution time of GAA2 is about half that of BGA and is about 0.05 times that of the SA-based algorithm. The SA algorithm is the slowest.

The solid curve in Fig. 9 shows the variation of the cost of the cheapest schedule found as the number of iterations increases in the GAA2 optimisation process. From this typical convergence characteristic in the figure, it can be observed that at the 125th iteration of GAA2, schedules in the near neighbourhood of the optimum schedule solution are already found. Between the 125th to 300th iteration, only a small improvement of the schedule solution is achieved and the schedule with the least cost up to the 300th iteration is taken as the optimum schedule. To account for the randomness in GAA2, it has been executed 30 times. The convergence characteristics of the different executions are bounded within the shaded area in Fig. 9. At the 300th iterations, the maximum deviation of the cost value in all the 30 executions is only 1.7609E-03%, confirming that the GAA2 is very reliable.

Fig. 10 summarises the typical convergence characteristics in 300 iterations of GAA2, GAA, BGA, IGA and SA when they are applied to the present application example. SA has the worst convergence characteristic amongst all the algorithms. IGA requires much less iterations than the others but, as mentioned earlier, it is prone to provide the worst solution among the GA-based algorithms. Similarly, BGA and GAA converge faster than GAA2 in terms of iteration number. However, the execution time in a BGA or GAA iteration is much longer than that of a GAA2 iteration as can be seen in Table 4. By considering both the computation time and the quality of the solution, GAA2 has the best performance.

9 CONCLUSION

After reporting a floating point number coding method for the representation of chromosomes to alleviate the discretisation problem intrinsic in the binary coding method, two hybrid optimisation algorithms, GAA and GAA2, have been developed based on genetic algorithms and the simulated annealing technique. By incorporating the methods developed into the hybrid algorithms, the premature convergence problem has been avoided and the adverse effects of the mutation operation has been reduced greatly. The performance of the hybrid algorithm GAA2 has been analysed and compared to other GA-based algorithms, SA-based algorithm and GAA. An expression serving as a guideline for setting the parameters for executing GAA2 when parameters settings for GAs and GAA are given has been derived.

The performance of the hybrid algorithms has been demonstrated through an application example in which the optimum generation schedule of a mixed hydro and thermal power system is sought. The study results and the convergence characteristics of the algorithms have shown that GAA2 is the most promising amongst all the algorithms considered.

Owing to the high computational requirement of the sequential forms of GAs and SA, the developed hybrid algorithms, though computationally much faster than pure SA and GAs, may still be regarded as inefficient for some practical applications. Parallel forms of the hybrid algorithms are reported in a companion paper [38] for reducing the computational requirement.

10 ACKNOWLEDGMENT

The work reported in this paper is supported by the Electricity Supply Association of Australia and by the Energy Research and Development Corporation. Their generous support is gratefully acknowledged.

11 REFERENCES

1. HOLLAND, J.H.: 'Adaptation in natural and artificial systems', (Ann Arbor: University of Michigan Press, 1975).
2. GOLDBERG, D.E.: 'Genetic algorithms in search, optimisation and machine learning' (Addison- Wesley, Reading, 1989).
3. BALA, J. W. and DE JONG K.: 'Generation of feature detectors for texture discrimination by genetic search', Proceedings of the 2nd International IEEE Conference on Tools for AI, 1990, pp. 812-818.
4. MANDAVA, V.R., FITZPATRICK M., and PICLENS, D. R.: 'Adaptive search space scaling in digital image registration', IEEE Transactions on Medical Imaging, 1989, 8(3), pp. 251-262.

5. COHOON, J.P., and PARIS, W. D.: 'Genetic Placement' IEEE Transactions on Computer-Aided Design, 1987, Vol 6 (6), pp. 956-964.

6. COHOON, J. P., HEDGE, S.U., MARTIN, W.N., and RICHARDS, D.S.: 'Distributed genetic algorithms for the floorplan design problem', IEEE Transactions on Computer-Aided Design, Vol , 1991, 10(4), pp. 483 - 492.

7. SHAHOOKAR K. and MAZUMDER, P.: 'A genetic approach to standard cell placement using meta-genetic parameter optimisation', IEEE Transactions on Computer-Aided Design, 1990, Vol. 9(5), pp. 500-511.

8. PARKER J. K. and Goldberg, D.E.: 'Inverse kinematics of redundant robots using genetic algorithm', Proceedings, IEEE International Conference on Robotics and Automation, 1989, pp. 271-276.

9. VIGNAUX, G.A. and MICHALEWICZ, Z: 'A genetic algorithm for the linear transportation problem', IEEE Transaction on Systems, Man and Cybernetics, 1989, Vol 21 (2), pp. 321-326.

10. THANGIAH, S.R., NYGARD, K.E. and JUELL, P. L.: 'Gideon: a genetic algorithm system for vehicle routing with time windows', Proceedings, 7th IEEE Conference on AI Applications, 1991, pp. 322-328.

11. WALTER, D.C. and SHEBLE, G.B.: 'Genetic algorithm solution of short term hydro-thermal scheduling with valve point loading', IEEE PES Summer Meeting, 1992, Seattle, Paper Number 92 SM 414-3 PWRS.

12. WONG, K.P., and WONG Y.W.: 'Genetic and genetic/simulated-annealing approaches to economic dispatch', to appear in IEE Proc. C, 1994.

13. YIN, X and GERMAY, N.: 'Investigations on solving the load flow problem by genetic algorithms', Electr Power Sys. Res., 1991, 22, pp. 151-163.

14. BISHOP, R.R. and RICHARDS, G.G.: 'Identifying induction machine parameters using a genetic opimization algorithm', IEEE Proceedings, Section 6C2, 1990, pp. 476-479.

15. NARA K., SATOH T. and KITAGAWA M.: 'Distribution systems loss minimum re-configuration by genetic algorithm', Conf. Proc. 3rd on Expert Systems Application to Power Systems, 1991, pp. 724-730.

16. MICHALEWICZ, Z.: 'Genetic algorithms + data structures = evolution programs' (Springer-Verlag, 1992).

17. MAULDIN, M.L.: 'Maintaining diversity in genetic search', AAAI Proc. National Conference on Artificial Intelligence, 1984, pp. 247-250.

18. GREFENSTETTE, J.J.: 'Optimization of control parameters for genetic algorithms', IEEE Transaction on Systems, Man and Cybernetics, 1986, vol 16(1), pp. 122-128.

19. TANESE, R. 'Parallel genetic algorithm for a hypercube', Proceedings of the 2nd International Conference on Genetic Algorithm, pp. 177-183.

20. BOOKER, L: 'Improving search in genetic algorithms', in Genetic Algorithms and Simulated Annealing, (Pitman, London, 1987), pp. 61-73.

21. De JONG, K.A.: 'An analysis of the behavior of a class of genetic adaptive systems' Doctoral Dissertation, University of Michigan, 1975.

22. KIRKPATRICK, S., GELATT, C.D., Jr., and VECCHI, M.P.: 'Optimisation by simulated annealing', Science, 1983, 220(4598), pp. 671-680.

23. AARTS, E., and KORST, J.M.: 'Simulated annealing and boltzmann machines: a stochastic approach to combinatorial optimisation and neural computing' (John Wiley, New York, 1989).

24. MUHLENBEIN, H. and KINDERMANN, J.: 'The dynamics of evolution and learning - towards genetic neural networks', in Connectionism in Perspective, (Elsevier Science Publishers B.V.), 1989, pp. 173-197.

25. HOLLAND, J.H., HOLYOAK, K.J., NISBETT, R.E. and THAGARD, P.R.: 'Classifier systems, Q-Morphisms, and induction', in Genetic Algorithms and Simulated Annealing, (Pitman, London, 1987), pp. 116-128.

26. GOLDBERG, D.E.: 'Computer-aided gas pipeline operation using genetic algorithms and rule learning', PhD thesis, University of Michigan, 1983.

27. FOGARTY, T.C.: 'An incremental genetic algorithm for real-time learning', Proc. 6th International Workshop on Machine Learning, 1989, Cornell, New York, pp. 416-419.

28. FOGARTY, T.C.: 'An incremental genetic algorithm for real-time optimisation', IEEE Conf. Proc. on Systems, Man and Cybernetics, 1989, pp. 321-326.

29. WONG, K.P., and WONG Y.W.: 'Floating-point number coding method for genetic algorithms', Conf. Proc. IEEE First Australian and New Zealand on Intelligent Information Systems, Perth, 1993, pp. 512-516.

30. LIN, F.T., KAO, C.Y. and HSU, C.C.: 'Applying the genetic approach to simulated annealing in solving some NP-hard prolems', IEEE Transaction on Systems, Man and Cybernetics, 1993, vol 23(6), pp. 1752-1767.

31. ESBENEN, H and MAZUNDER, P.: 'SAGA: a unification of the genetic algorithm with simulated annealing and its application to macro-cell placement', 7th Interantional Conference on VLSI design, 1994., pp. 221-214.

32. SZU, H. and HARTLEY, R.: 'Fast simulated annealing', Physics Letters A, 122, 1987, pp. 157- 162.

33. WONG, K.P. and FUNG, C.C.: 'Simulated-annealing-based economic dispatch algorithm', IEE Proc. C, vol. 140, no. 6, Nov. 1993, pp. 509-515.

34. WILSON, S.W., 'Classifier systems and the animate problem', Research Memo RIS-27r, Rowland Institute for Science, Cambridge.

35. MAULDIN, M.L.: 'Maintaining diversity in genetic search', AAAI Proc. National Conference on Artificial Intelligence, 1984, pp. 247-250

36. WOOD, A.J. and WOLLENBERG, B.F.: 'Power Generation, Operation and Control' (Wiley, New York, 1984).

37. WONG, K.P., and WONG Y.W.: 'Short-term hydrothermal-scheduling: Part 1 simulated annealing approach', to appear in IEE Proc. C, 1994.

38. WONG, K.P., and WONG, Y.W.: 'Development of parallel hybrid optimisation techniques based on Genetic Algorithms and Simulated Annealing', companion paper in Proc. of AI'94 Workshop on Evolutionary Computation, Armidale, Australia, Nov. 1994.

Table 2. Determined hydrothermal schedules and fuel costs

method	interval	demand (MW)	therm gen (MW)	hydro gen (MW)	discharge (acre-ft/hr)	volume (acre-ft)	cost
SA	1st day 1st int.	1200	893.73	306.27	1852.18	101773.81	
	2nd int.	1500	895.24	604.76	3335.65	85746.05	
	2nd day 1st int.	1100	884.32	215.68	1401.93	92922.88	
	2nd int.	1800	912.22	887.78	4742.27	60015.62	
	3rd day 1st int.	950	781.87	168.13	1165.60	70028.41	
	2nd int.	1300	795.83	504.17	2835.70	60000.01	709874.36
BGA	1st day 1st int.	1200	893.41	306.59	1853.74	101755.09	
	2nd int.	1500	897.25	602.75	3325.66	85847.20	
	2nd day 1st int.	1100	896.24	203.76	1342.71	93734.70	
	2nd int.	1800	898.35	901.65	4811.19	60000.4	
	3rd day 1st int.	950	787.86	162.14	1135.81	70370.6	
	2nd int.	1300	790.10	509.90	2864.22	60000.0	709862.40
IGA	1st day 1st int.	1200	896.30	303.70	1839.38	101927.47	
	2nd int.	1500	896.92	603.08	3327.31	85999.73	
	2nd day 1st int.	1100	896.29	203.71	1342.45	93890.33	
	2nd int.	1800	895.74	904.26	4824.16	60000.44	
	3rd day 1st int.	950	788.01	161.99	1135.11	70379.08	
	2nd int.	1300	789.96	510.04	2864.92	60000.00	709862.05
GAA	1st day 1st int.	1200	897.85	302.15	1831.69	102019.75	
	2nd int.	1500	895.03	604.97	3336.71	85979.19	
	2nd day 1st int.	1100	896.26	203.74	1342.59	93868.16	
	2nd int.	1800	896.11	903.89	4822.34	60000.08	
	3rd day 1st int.	950	787.84	162.16	1135.92	70369.08	
	2nd int.	1300	790.12	509.88	2864.09	60000.00	709862.17
GAA2	1st day 1st int.	1200	895.45	304.55	1843.61	101876.63	
	2nd int.	1500	896.16	603.84	3331.07	85903.78	
	2nd day 1st int.	1100	896.13	203.87	1343.22	93785.09	
	2nd int.	1800	897.51	902.49	4815.40	60000.31	
	3rd day 1st int.	950	789.08	160.92	1129.78	70442.95	
	2nd int.	1300	788.88	511.12	2870.25	60000.00	709862.086
gradient (Ref 36)	1st day 1st int.	1200	903.11	296.89	1805.56	102333.3	
	2nd int.	1500	889.22	610.78	3365.56	85946.7	
	2nd day 1st int.	1100	893.65	206.35	1355.56	93680.0	
	2nd int.	1800	899.26	900.74	4806.67	60000.0	
	3rd day 1st int.	950	806.25	143.75	1044.44	71466.7	
	2nd int.	1300	771.72	528.28	2955.56	60000.1	709877.38

Table 3. Comparison of worst fuel costs

algorithm	best cost	worst cost	% increment in cost
gradient	709877.38		
SA	709874.36	710717.07	0.12
BGA	709862.40	709992.35	0.0162
IGA	709862.05	709899.69	0.0031
GAA	709862.17	709869.17	0.00
GAA2	709862.09	709865.51	0.00

Table 4. Execution times of the new algorithms and SA

execution time (sec.)	SA	BGA	IGA	GAA	GAA2
shortest	891.77	42.28	113.80	116.95	40.38
longest	917.28	104.33	137.92	146.71	47.93
average	901.44	85.77	120.15	121.02	45.16

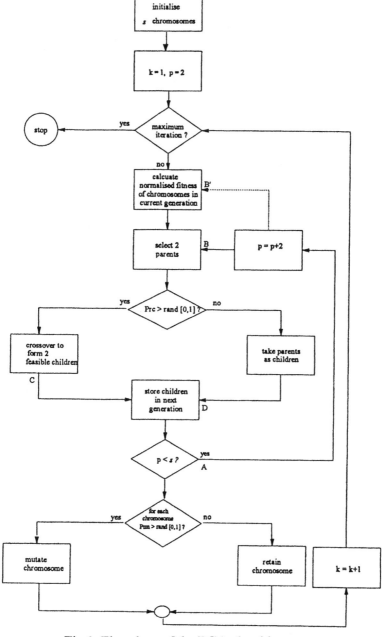

Fig.1: Flowchart of the BGA algorithm

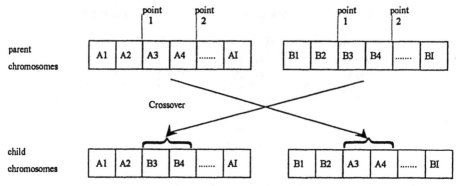

Fig.2: Representation of chromosomes and 2-point crossover operation

Ai and Bi are values of parameter i. Values are represented directly by floating-point numbers

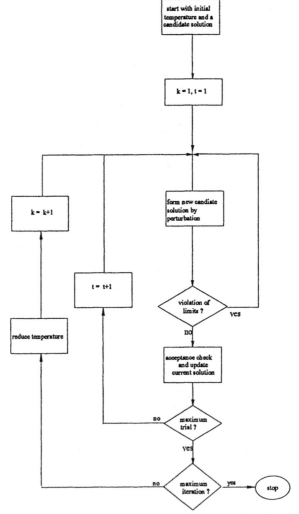

Fig.3: Flow chart of the SA algorithm

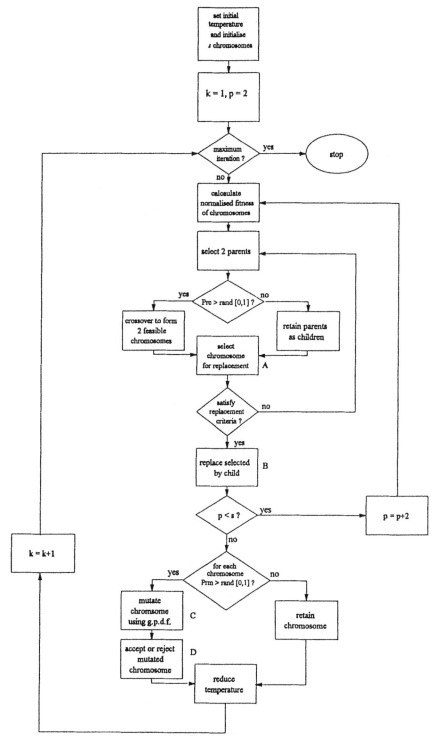

Fig.4: Flowchart of the GAA algorithm

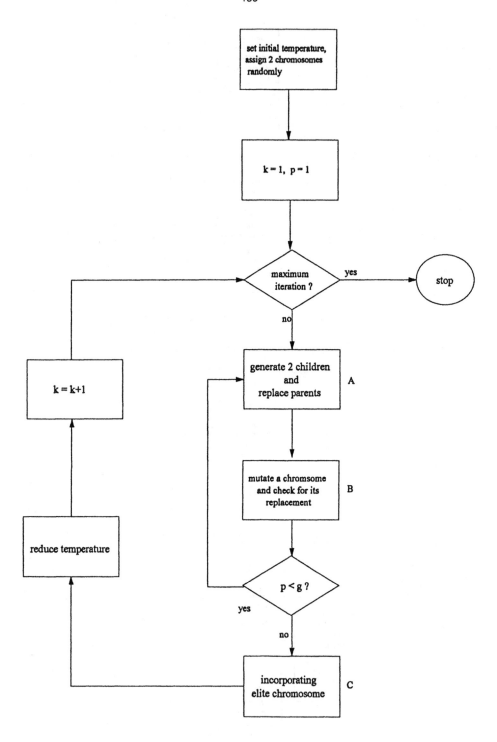

Fig.5: Flowchart of the GAA2 algorithm

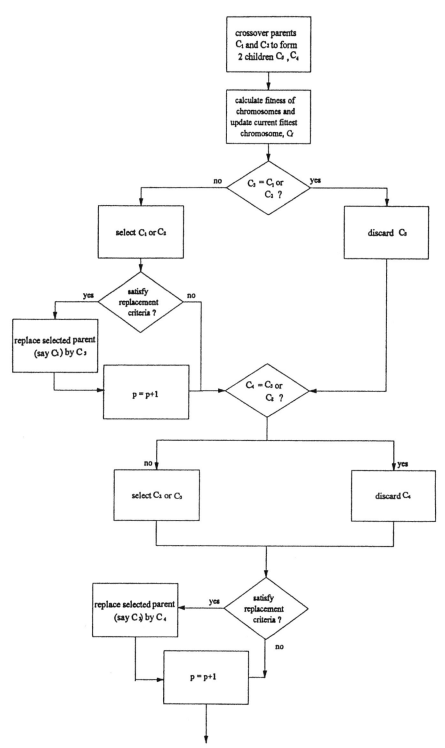

Fig.6: Flowchart of block A in Fig.5

152

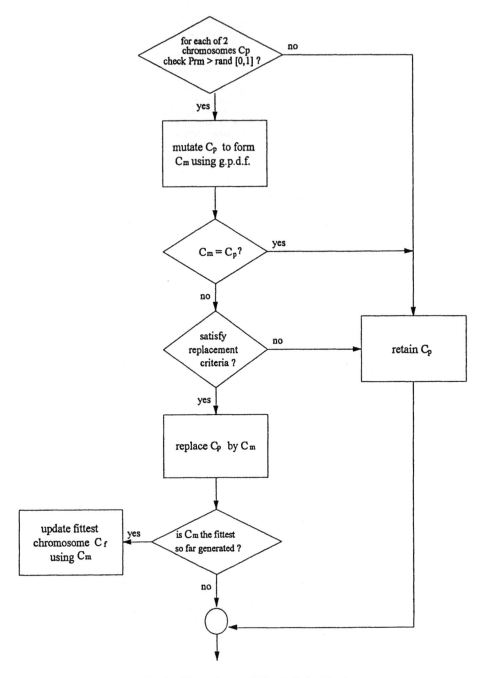

Fig.7: Flowchart of block B in Fig.5

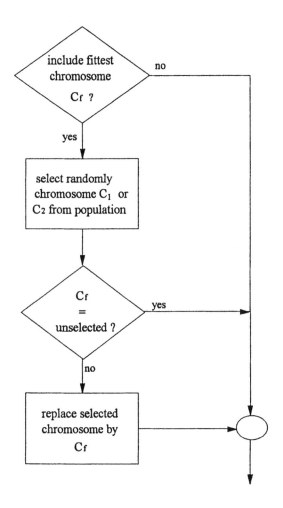

Fig. 8: Flowchart of block C in Fig.5

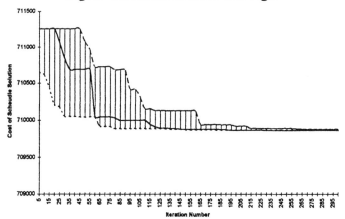

Fig. 9: Convergence characteristics of GAA2

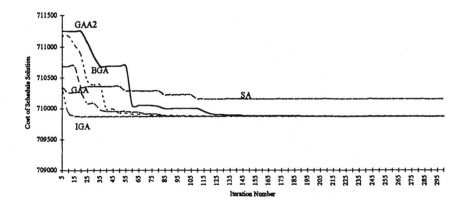

Fig. 10 Comparison of convergence characteristics of GAA2, GAA, BGA, IGA and SA

Development of Parallel Hybrid Optimisation Techniques Based on Genetic Algorithms and Simulated Annealing

Kit Po Wong and Yin Wa Wong

Artificial Intelligence and Power Systems Research Group, Department of Electrical and Electronic Engineering, The University of Western Australia, Nedlands, Western Australia 6009

Abstract. This paper develops coarse-grained parallel algorithms for hybrid optimisation techniques based on genetic algorithms and simulated annealing. The design of the algorithms takes into consideration load balancing, processor synchronisation reduction, communication overhead reduction and memory contention elimination. In addition, the algorithms are designed to avoid the problem of premature convergence which exists in some previous parallel genetic algorithms. The algorithms are implemented on an i860 processor in a simulated environment and are applied to a short-term hydrothermal scheduling problem. The scheduling results are presented and are compared to those found by sequential GAs, by parallel simulated-annealing algorithm and by some earlier parallel genetic algorithms.

1 INTRODUCTION

In the companion paper [1], sequential hybrid algorithms GAA and GAA2 based on genetic algorithms (GAs) [2, 3] and simulated-annealing (SA) [4, 5] have been developed for solving optimisation problems. Owing to the high computational requirement of the sequential forms of GAs and SA [6-8], the developed hybrid algorithms, though computationally much faster than pure SA and GAs, may not be fast enough for some practical applications.

Parallel processing is one of the promising means to increase the computing speed of the hybrid algorithms. Processors such as i860's are powerful and high-speed numerical processors, which can work with PC-486 computers. With the availability of software systems to operate these processors in parallel in the near future, it can be envisaged that the hybrid algorithms in the companion paper [1] can be developed further to exploit parallelism. In general, parallel algorithms should be designed with the considerations of (a) the preservation of the basic requirements in the sequential algorithm, (b) the hardware architecture of the processors, (c) the configurations of the processors and (d) the software environment of the processors. Items (b)- (d), in addition, are particularly relevant to the type of processors adopted.

A parallel algorithm for the basic genetic algorithms, PBGA [9] have recently been developed by the authors. In its development, the considerations of all of the factors affecting the design of parallel algorithms were taken into account. Compared to PBGAs reported previously [10-12], the authors' PBGA has the desirable feature that it is able to reduce the chance of occurrence of premature convergence.

Based on the PBGA established by the authors, this paper develops coarse-grained [13] parallel algorithms for the hybrid optimisation techniques GAA and GAA2. A parallel form of incremental genetic algorithms (IGA) is also developed since the GAA and GAA2 algorithms have IGA as their basis. In the development of the parallel algorithms, the design factors in items (a) - (d) above are fully considered. The algorithms are implemented in a simulated parallel environment on an i860 processor.

An i860 processor is a Single Function Multiple Data processor and belongs to the class of coarse-grained processors. The i860 processors have local memories and they can execute identical function modules in parallel with different input data. These parallel processors communicate with the host through one data channel only.

The performance of the parallel hybrid algorithms is demonstrated through an application example. In the application example, the short-term hydrothermal scheduling problem investigated in the companion paper [1] is adopted.

2   A PARALLEL BGA

The performance of GAs depends on the diversity of information held in the chromosomes. Since diversity usually decreases with decreasing population size, the performance of the GAs solution process will deteriorate with a smaller population size and will converge to a local optimum prematurely. In the sequential BGA, a chromosome is selected for the crossover operation based on its fitness over the entire population. In the PBGAs previously proposed by Pettey [10], Cohoon [11] and Tanese [12], a chromosome is selected based on its fitness in the local subpopulation of the processor. The subpopulation is of much smaller size than the whole population and this can lead to premature convergence.

To overcome the premature convergence problem, a new PBGA has been proposed by the authors [9]. In the new PBGA, the total population size p is assigned to each of a total of I processors instead of p/I, which is assigned to the previous PBGAs by Pettey [10], Cohoon [11] and Tanese [12]. The authors' PBGA is summarised below.

1. In the initialisation process, while all I processors are processing in parallel, p chromosomes are initialised independently in each processor.
2. In the production of a generation of chromosomes, each processor disjointedly and in parallel, executes a GA on its own population of p chromosomes. However, instead of producing p chromosomes, as in the sequential GAs, each processor produces only p/I chromosomes, where I is the number of processors. This is shown in Fig. 1.

3. At the end of each generation, each processor passes its new generation, which consists of p/I chromosomes, to other processors through the host in the following manner:

   (a) Only some of the elements in a chromosome are passed to another processor. Based on the passed elements, the other processors need to re-establish the omitted elements and recalculate the fitness value of the chromosome.

   (b) In the late stage of the GAs solution process, most chromosomes converge to an identical form. Based on this, in the late stage only the number and the form of the identical chromosomes are passed to other processors, which will then re-establish all the identical chromosomes.

   Refinements in items (a) and (b) can greatly reduce the communication overhead and the memory contention problem, which are intrinsic in the transference of all the elements and the fitness value of the p/I new chromosomes from a processor to the host, then to other processors.

4. Steps 2 and 3 are repeated until the maximum allowable number of generations is reached. Alternatively, the process is terminated when a chromosome in one of the processors has achieved the desired fitness.

Since all the processors execute a GA on p chromosomes to generate identical numbers of chromosomes, they are loaded evenly and consume almost the same execution time and hence the effort to synchronise the processors at the completion of a generation is small.

3 PARALLEL IGA AND PARALLEL GAA

The PBGA in Section (2) can be adopted easily to establish a parallel form of IGA and a parallel form of GAA. A minor modification is first made in step (2). In that step, an IGA or a GAA is executed instead of a GA.

In IGA and GAA, the child chromosomes produced by two selected parents are included in the current generation instead of in the next generation. Therefore, there is no distinction between the next generation and the current generation. It follows that to implement the parallel IGA (PIGA) and parallel GAA (PGAA), a newly produced chromosome is stored in readiness for passing to the host at the end of each generation in addition to being included in the current generation.

4 PARALLEL GAA2

Due to the much smaller pseudo-population size, g, and the larger number of generations or iterations in GAA2, PBGA in Section (2) is modified to establish a parallel GAA2 (PGAA2). The algorithm is summarised below and is also shown in Fig. 2.

1. In the initialisation of the whole population, each of a total of I processors works in parallel and disjointedly to initialise 2 chromosomes.

2. At each generation, each processor, disjointedly and in parallel, executes a GAA2 on its population. Instead of producing g chromosomes according to the pseudo population size of the sequential GAA2, each processor produces only g/I chromosomes. The pseudo subpopulation size is g/I.

3. At the end of each predetermined number of generations, m, each processor passes to the host the 2 chromosomes in the mth generation and the fittest chromosome generated so far. Fig. 2 shows this for the case of the 2nd processor. In the figure, Pi denotes one of the two chromosomes in the mth generation and Fi denotes the fittest chromosome generated so far.

   The host will pass the fittest chromosome (F) among the Fi's of all the processors to a randomly chosen processor (R). The selection of the 2nd chromosome to be passed to processor R and the selection of any one of the 2 chromosomes to be passed to the other processors are implemented in the following manner:

   A binary digit is randomly generated first. If its value is 1, an arbitrary chosen chromosome from the set of Fi will be passed. If its value is 0, a randomly chosen Pi will be passed.

   For processor R, before the selected chromosome is passed to it from the host, the chromosome is checked to ensure that the chromosome is distinct from the chromosome already in R. If the chromosome is not distinct, it is discarded and another chromosome will be selected by the method above. For the passing of chromosomes from the host to the other processors, the distinct chromosome checking will be performed before the second chromosome is passed.

4. The procedure consisting of steps (2) and (3) is repeated until the maximum number of generations has been reached or a chromosome in one of the processors has achieved the desired fitness.

In the above implementation, since information is passed from a processor to the host every m generations instead of every generation, the communication and synchronisation effort will be reduced. However, as the number of processors I increases, the pseudo subpopulation size, g/I , of each processor will decrease. To alleviate the adverse effects due to the small pseudo subpopulation size, the frequency of information passing should be increased. The value of number m can be modified according to K/I where K is a positive integer number and is a multiple of I. In the application example in the next section, K is set to 16 and I is 4. The reduction of the value of m implies that the frequency of information passing will be increased. Provided that K is not equal to I, the communication and synchronisation effort will always be reduced.

5 APPLICATION EXAMPLE

The PBGA, PIGA, PGAA and PGAA2 have been applied to solve the hydro-thermal scheduling problem in the companion paper [1]. These algorithms have been implemented using C in a simulated parallel environment and are run on an i860 numerical processor.

Four i860 processors have been simulated for the execution of the parallel algorithms. The parameter settings for executing the parallel scheduling algorithms are identical to those used in the sequential algorithms [1]. As in the case of the sequential algorithms, the parallel algorithms are executed 30 times.

Table 1 summarises the best hydrothermal schedules found by the sequential algorithms reported in the companion paper [1]. In this table and Tables 2 and 4, '1st int.' is the first interval which starts from 12.00 am and ends at 12.00 pm. '2nd int.' is the second interval which starts from 12.00 pm and ends at 12.00 am. Table 2 tabulates the best schedules obtained by parallel algorithms of BGA, IGA and SA together with those obtained by earlier parallel BGAs due to Pettey [10], Cohoon [11] and Tanese [12]. From the results in Tables 1 and 2, it can be observed that the qualities of solution schedules by genetic-based algorithms both in sequential and in parallel forms are comparable except Pettey's. Although parallel SA performs better than its sequential counterpart and Pettey's PBGA, it is inferior to the other algorithms considered.

Table 3 summarises the worst schedule solutions found by all the algorithms. The results show that previous parallel algorithms and the parallel form of SA by the authors are less reliable than the parallel BGA developed in Section 2.

Table 4 tabulates the best schedules found by the sequential and parallel hybrid algorithms GAA and GAA2. The schedules are almost identical but PGAA2 has the best result. PGAA2's schedule is also the best among all the other algorithms. The 'worst' results in Table 3 also indicate that PGAA2 can find the best 'worst' solution schedule.

By examining the summary of the CPU times required by all the algorithms in Table 5, it can be observed that PGAA2 is the fastest. This shows that PGAA2 has good performance with respect to the quality of solution and the computing time requirement. From the results in Table 5, the average cpu time of the GAA and GAA2 is 3.69 times that of the PGAA and PGAA2 respectively. While the ideal speed gain factor is 4 because four i860 processors have been simulated in this application example, a near linear gain in speed has been achieved by the developed PGAA and PGAA2.

6 CONCLUSION

Coarse-grained parallel algorithms have been developed for the hybrid optimisation techniques GAA and GAA2 based on a parallel form of BGA designed by the authors [9]. A parallel form of IGA has also been developed. The design of the parallel algorithms has taken into consideration load balancing, processor synchronisation reduction, communication overhead reduction and memory contention elimination.

Using the short-term hydrothermal scheduling problem in the companion paper [1] as the test example for the developed algorithms, it has been shown

Table 1. Best schedules obtained by the sequential algorithms

method	interval	demand (MW)	therm gen (MW)	hydro gen (MW)	discharge (acre-ft/hr)	volume (acre-ft)	cost
SA	1st day 1st int.	1200	893.73	306.27	1852.18	101773.81	
	2nd int.	1500	895.24	604.76	3335.65	85746.05	
	2nd day 1st int.	1100	884.32	215.68	1401.93	92922.88	
	2nd int.	1800	912.22	887.78	4742.27	60015.62	
	3rd day 1st int.	950	781.87	168.13	1165.60	70028.41	
	2nd int.	1300	795.83	504.17	2835.70	60000.01	709874.36
BGA	1st day 1st int.	1200	893.41	306.59	1853.74	101755.09	
	2nd int.	1500	897.25	602.75	3325.66	85847.20	
	2nd day 1st int.	1100	896.24	203.76	1342.71	93734.70	
	2nd int.	1800	898.35	901.65	4811.19	60000.4	
	3rd day 1st int.	950	787.86	162.14	1135.81	70370.6	
	2nd int.	1300	790.10	509.90	2864.22	60000.0	709862.40
IGA	1st day 1st int.	1200	896.30	303.70	1839.38	101927.47	
	2nd int.	1500	896.92	603.08	3327.31	85999.73	
	2nd day 1st int.	1100	896.29	203.71	1342.45	93890.33	
	2nd int.	1800	895.74	904.26	4824.16	60000.44	
	3rd day 1st int.	950	788.01	161.99	1135.11	70379.08	
	2nd int.	1300	789.96	510.04	2864.92	60000.00	709862.05
GAA	1st day 1st int.	1200	897.85	302.15	1831.69	102019.75	
	2nd int.	1500	895.03	604.97	3336.71	85979.19	
	2nd day 1st int.	1100	896.26	203.74	1342.59	93868.16	
	2nd int.	1800	896.11	903.89	4822.34	60000.08	
	3rd day 1st int.	950	787.84	162.16	1135.92	70369.08	
	2nd int.	1300	790.12	509.88	2864.09	60000.00	709862.17
GAA2	1st day 1st int.	1200	895.45	304.55	1843.61	101876.63	
	2nd int.	1500	896.16	603.84	3331.07	85903.78	
	2nd day 1st int.	1100	896.13	203.87	1343.22	93785.09	
	2nd int.	1800	897.51	902.49	4815.40	60000.31	
	3rd day 1st int.	950	789.08	160.92	1129.78	70442.95	
	2nd int.	1300	788.88	511.12	2870.25	60000.00	709862.086

that the qualities of the solutions of the developed parallel GAs and the parallel hybrid algorithms are comparable to those of their sequential forms. However, the developed parallel algorithms are much faster and a near-linear reduction in speed has been achieved when compared to the sequential algorithms. Among all the parallel algorithms, PGAA2 is the fastest and is also the most reliable in terms of solution quality.

The ideal gain in speed has not been achieved by the parallel algorithms owing to the times required by (a) the communication between the host and the processors, (b) process creation and termination in the i860s and (c) the synchronisation of the processors. Moreover, some components of the parallel algorithms, such as the selections of population for each of the processors in the PGAA2, cannot be executed in parallel.

7 ACKNOWLEDGMENT

The generous support given to the work reported in this paper by the Energy Research and Development Corporation and by the Electricity Supply Association of Australia is gratefully acknowledged.

8 REFERENCES

1. WONG, K.P., and WONG, Y.W.: 'Development of hybrid optimisation techniques based on Genetic Algorithms and Simulated Annealing', companion paper in Proc. of AI'94 Workshop on Evolutionary Computation, Armidale, Australia, Nov. 1994.
2. HOLLAND, J.H.: 'Adaptation in natural and artificial systems', (Ann Arbor: University of Michigan Press, 1975)
3. GOLDBERG, D.E.: 'Genetic algorithms in search, optimisation and machine learning' (Addison-Wesley, Reading, 1989)
4. KIRKPATRICK, S., GELATT, C.D., Jr., and VECCHI, M.P.: 'Optimisation by simulated annealing', Science, 1983, 220(4598), pp. 671-680.
5. AARTS, E., and KORST, J.M.: 'Simulated annealing and boltzmann machines: a stochastic approach to combinatorial optimisation and neural computing' (John Wiley, New York, 1989).
6. SANNIER, A.V. and GOODMAN, E.D.: 'Genetic learning procedures in distributed environments', Proceedings of the 2nd International Conference on Genetic Algorithm, pp. 162-169.
7. VIGNAUX, G.A. and MICHALEWICZ, Z: 'A genetic algorithm for the linear transportation problem', IEEE Transactions on Systems, Man and Cybernetics, 1989, Vol 21 (2), pp. 321-326.
8. KADABA, N. and NYGARD, K.E.:, 'Improving the performance of genetic algorithms in automated discovery of parameters', Proceedings of the Seventh International Conference on Machine Learning, 1990, pp. 140-148.
9. WONG, K.P., and WONG, Y.W.: 'Development of parallel genetic algorithms' Australian Journal of Intelligent Information Processing Systems, 1994, Vol. 1, (1) pp. 51-57.
10. PETTEY, C.B., LEUZE, M.R. and GREFENSTETTE, J. J: 'A parallel genetic algorithm', Proceedings of the 2nd International Conference on Genetic Algorithm, 1987, pp. 155-161.
11. COHOON, J. P., HEDGE, S.U., MARTIN, W.N., and RICHARDS, D.S.: 'Distributed genetic algorithms for the floorplan design problem', IEEE Transactions on Computer Design, Vol , 1991, 10(4), pp. 483 - 492.
12. TANESE, R.: 'Parallel genetic algorithm for a hypercube', Proceedings of the 2nd International Conference on Genetic Algorithm, 1987, pp. 177-183.
13. IEEE Committee Report: 'Parallel processing in power systems computation' IEE/PES , Summer Meeting 1991, Paper Number 91 SM 503-3 PWRS.

Table 2. Best schedules obtained by parallel BGA, IGA and SA

method	interval	demand (MW)	therm gen (MW)	hydro gen (MW)	discharge (acre-ft/hr)	volume (acre-ft)	cost
Cohoon	1st day 1st int.	1200	896.64	303.36	1837.71	101947.50	
	2nd int.	1500	896.20	603.80	3330.88	85977.00	
	2nd day 1st int.	1100	895.60	204.40	1345.89	93826.33	
	2nd int.	1800	896.84	903.16	4818.69	60002.02	
	3rd day 1st int.	950	791.16	158.84	1119.44	70568.70	
	2nd int.	1300	786.78	513.22	2880.73	60000.00	709862.42
Pettey	1st day 1st int.	1200	894.65	305.35	1847.60	101828.81	
	2nd int.	1500	898.28	601.72	3320.53	85982.47	
	2nd day 1st int.	1100	894.93	205.07	1349.18	93792.28	
	2nd int.	1800	897.57	902.43	4815.08	60011.30	
	3rd day 1st int.	950	767.55	182.45	1236.76	69170.18	
	2nd int.	1300	810.22	489.78	2764.18	60000.00	709883.18
Tanese	1st day 1st int.	1200	899.83	300.17	1821.86	102137.69	
	2nd int.	1500	893.56	606.44	3344.01	86009.53	
	2nd day 1st int.	1100	897.67	202.33	1335.56	93982.83	
	2nd int.	1800	894.46	905.54	4830.55	60016.24	
	3rd day 1st int.	950	789.04	160.96	1129.98	70456.48	
	2nd int.	1300	788.66	511.34	2871.37	60000.00	709863.88
PBGA	1st day 1st int.	1200	896.70	303.30	1837.42	101951.02	
	2nd int.	1500	896.43	603.57	3329.77	85993.81	
	2nd day 1st int.	1100	896.54	203.46	1341.20	93899.38	
	2nd int.	1800	895.69	904.31	4824.41	60006.42	
	3rd day 1st int.	950	788.38	161.62	1133.23	70407.69	
	2nd int.	1300	789.48	510.52	2867.31	60000.00	709862.55
PIGA	1st day 1st int.	1200	896.72	303.28	1837.32	101952.16	
	2nd int.	1500	896.14	603.86	3331.19	85977.92	
	2nd day 1st int.	1100	895.98	204.02	1343.96	93850.36	
	2nd int.	1800	896.41	903.59	4820.85	60000.16	
	3rd day 1st int.	950	786.61	163.39	1142.03	70295.78	
	2nd int.	1300	791.35	508.65	2857.98	60000.00	709862.22
PSA	1st day 1st int.	1200	905.09	294.91	1795.72	102451.36	
	2nd int.	1500	888.98	611.02	3366.79	86049.91	
	2nd day 1st int.	1100	891.17	208.83	1367.87	93635.48	
	2nd int.	1800	900.89	899.11	4798.58	60052.52	
	3rd day 1st int.	950	791.17	158.83	1119.38	70620.00	
	2nd int.	1300	785.92	514.08	2885.00	60000.00	709870.46

Table 3. Costs of the best and worst schedules

method	best cost	worst cost
BGA	709,862.40	709,992.35
Cohoon	709,862.42	709,980.53
Pettey	709,883.18	710,023.76
Tanese	709,863.88	710,023.72
PBGA	709,862.55	709,877.35
IGA	709,862.05	709,899.69
PIGA	709,862.22	709,870.48
GAA	709,862.17	709,869.17
PGAA	709,862.44	709,877.16
GAA2	709,862.09	709,865.51
PGAA2	709,862.05	709,862.06
SA	709,874.36	710,717.07
PSA	709,870.46	710,023.77

Table 4. Determined hydrothermal schedules and fuel cost by the hybrid algorithms

method	interval	demand (MW)	therm gen (MW)	hydro gen (MW)	discharge (acre-ft/hr)	volume (acre-ft)	cost
GAA	1st day 1st int.	1200	897.85	302.15	1831.69	102019.75	
	2nd int.	1500	895.03	604.97	3336.71	85979.19	
	2nd day 1st int.	1100	896.26	203.74	1342.59	93868.16	
	2nd int.	1800	896.11	903.89	4822.34	60000.08	
	3rd day 1st int.	950	787.84	162.16	1135.92	70369.08	
	2nd int.	1300	790.12	509.88	2864.09	60000.00	709862.17
PGAA	1st day 1st int.	1200	895.09	304.91	1845.40	101855.23	
	2nd int.	1500	897.25	602.75	3325.66	85947.34	
	2nd day 1st int.	1100	895.63	204.37	1345.74	93798.50	
	2nd int.	1800	897.34	902.66	4816.21	60003.94	
	3rd day 1st int.	950	788.59	161.41	1132.20	70417.59	
	2nd int.	1300	789.31	510.69	2868.13	60000.00	709862.44
GAA2	1st day 1st int.	1200	895.45	304.55	1843.61	101876.63	
	2nd int.	1500	896.16	603.84	3331.07	85903.78	
	2nd day 1st int.	1100	896.13	203.87	1343.22	93785.09	
	2nd int.	1800	897.51	902.49	4815.40	60000.31	
	3rd day 1st int.	950	789.08	160.92	1129.78	70442.95	
	2nd int.	1300	788.88	511.12	2870.25	60000.00	709862.09
PGAA2	1st day 1st int.	1200	896.30	303.70	1839.38	101927.47	
	2nd int.	1500	896.92	603.08	3327.31	85999.73	
	2nd day 1st int.	1100	896.29	203.71	1342.45	93890.33	
	2nd int.	1800	895.74	904.26	4824.16	60000.44	
	3rd day 1st int.	950	788.01	161.99	1135.11	70379.08	
	2nd int.	1300	789.96	510.04	2864.92	60000.00	709862.05

Table 5. Maximum, minimum and average CPU times

method	shortest time (sec)	longest time (sec)	average time (sec)
SA	891.77	917.28	901.44
PSA	233.95	245.50	238.95
BGA	42.28	104.33	85.77
PBGA	30.67	32.90	31.68
IGA	113.80	137.92	120.15
PIGA	32.18	41.39	39.96
GAA	116.95	146.71	121.02
PGAA	31.80	33.98	32.76
GAA2	40.38	47.93	45.16
PGAA2	11.39	12.93	12.23

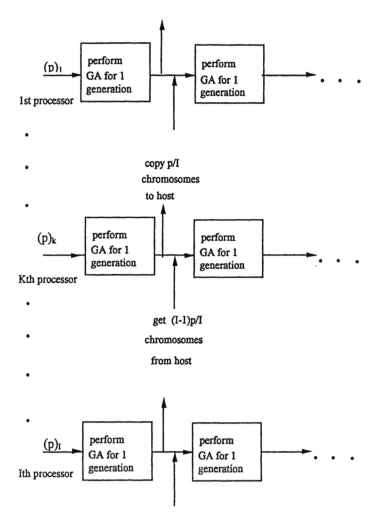

Fig. 1. Proposed PBGA

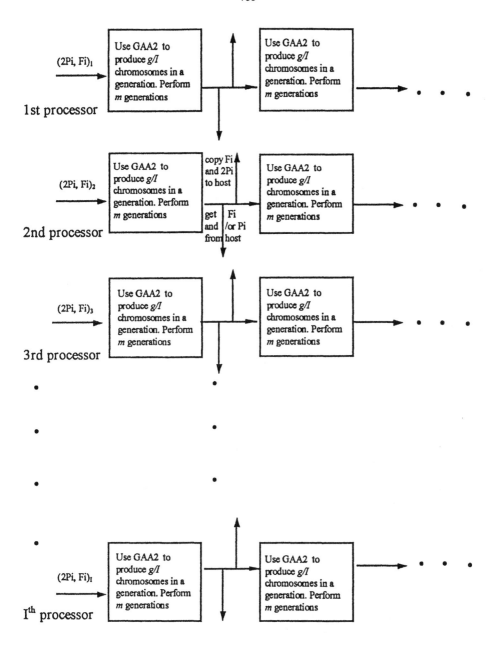

Fig. 2   Proposed PGAA2

Pi - one of the two chromosomes in the mth generation
Fi - the fittest chromosome generated so far.

Genetic Algorithms
for Cutting Stock Problems:
with and without Contiguity

Robert Hinterding and Lutfar Khan

Department of Computer and Mathematical Sciences
Victoria University of Technology
PO Box 14428 MMC,Melbourne 3000
email: rhh@matilda.vut.edu.au

Abstract. A number of optimisation problems involve the optimal grouping of a finite set of items into a number of categories subject to one or more constraints. Such problems raise interesting issues in mapping solutions in genetic algorithms. These problems range from the knapsack problem to bin packing and cutting stock problems. This paper describes research involving cutting stock problems. Results show that the mapping that is used affects the solution in terms of both quality of the solution found and time taken to find solutions, and that different mappings are suitable for different variants of the problem.

1 Introduction

Genetic algorithms(GAs) have been applied to a number of ordering problems, ranging from the Travelling Salesperson Problem (Goldberg 1985) to Job-shop Scheduling (Davis 1991). Typically these problems have been mapped by representing the solutions as ordered lists of alleles, each allele representing an item. Falkenauer (1991) has defined a sub-set of these problems as grouping problems. Such problems require the grouping of items into sets. An example is the classical Bin Packing Problem (Coffman et al 1984) where a finite set of items is to be packed into a minimum number of bins of the same size.

In this paper we look at a variant of the bin-packing problem; the problem of packing a finite set of items into a number of categories of different sizes. Such problems are common in real-life in the areas of time-tabling, stock-cutting and pallet loading. The problem chosen is that of cutting a set number of rods from stock lengths of different sizes such that wastage is minimised, called the Cutting Stock Problem(CSP).

In representing solutions to such problems, the standard representation of the solution as a permutated list has a number of problems. Items to be cut may be represented as a queue in a permutated list, but the problem remains as to how to represent the different sized stock sizes (categories). Typically GAs have coped with this problem by using an intelligent decoder (builder) to assign items to categories. For example, Syswerda (1991) uses an intelligent decoder or scheduler to assign items (jobs) from a queue (represented as a chromosome string) for a job-shop scheduling problem. Another approach is to use more than one chromosome per individual (Juliff 1993; Hinterding & Juliff 1993). Different aspects of a solution are encoded onto separate chromosomes and a decoder builds solutions, taking specifications from the separate chromosomes.

Another problem is that when chromosomes are used to represent groups of items, crossover disrupts the groups and convergence to a solution can be very slow. Falkenauer (1991) uses a "grouping" chromosome to overcome this problem. Here each gene represents a group of items, and hence crossover works with groups of items rather than a list of single items.

In this paper we build on earlier work (Hinterding & Juliff 1993; Hinterding 1994) to develop GAs for the CSP using two different mappings. Hinterding & Juliff implemented a multi-chromosome GA for cutting stock, and we dramatically improve and extend on their work. Hinterding (1994) explored two different mappings for solving the Knapsack problem. We extend the use of these mappings to the cutting stock problem. Falkenauer (1994) has extended his bin packing GA by using local optimisation to improve his results dramatically and we compare our GAs to his.

We restrict ourselves to using GAs with only one chromosome per individual, and by not using local optimisation. We do this to see the effects of the different mapping on the GAs. Problems tested range from 20 to 126 items with one to eight different stock sizes.

2 The Cutting Stock Problem

The cutting stock problem (CSP) is to minimise:

$$W = \sum_{j \in J} w_j x_j \tag{1}$$

subject to

$$\sum_{j \in J} a_{ij} x_j = N_i \quad \text{for} \quad i = 1, 2, \ldots, n \tag{2}$$

and

$$\text{for each} \quad j \in J, \quad x_j \text{ is a positive integer,} \tag{3}$$

where,

n = number of orders,

w_j = waste per run of pattern j,

a_{ij} = number of pieces of item i in pattern j,

x_j = number of runs of pattern j,

N_i = number of pieces of item i.

If there is only one stock length L, and l_i is the length of order i, then

$$L = \sum_{i=1}^{n} a_{ij} l_i + w_j, \text{ for all } j \in J,$$

and then CSP can be written as, minimise:

$$X = \sum_{j \in J} x_j$$

subject to (2) and (3). Note that

$$W = \sum_{j \in J} w_j x_j = X.L - \sum_{i=1}^{n} N_i l_i.$$

The cutting stock problem is one of the first decision problems of Operations Research modelled in a mathematical programming framework. This model and its variants have been widely used in the paper industry; paper is produced in standard lengths and then cut into appropriate sizes to meet customers' demands. Other application areas include steel mills and cable industries. When the items to be produced vary in one dimension (length), the problem is called 1-dimensional CSP. Item sizes specified in two or more dimensions have similarly been modelled. If there is more than one stock length from which the orders are cut, the problem is called multiple stock length CSP. A general description and classification of cutting problems is given in Dyckhoff (1990).

The pioneering work in solving a CSP was by Gilmore and Gomory (1963l). They used a linear programming model to solve it; the integrality constraints were relaxed initially to $x_j \geq 0$, and the LP solution was obtained by a clever method of column generation. To obtain an integer solution from the LP solution, usually simple rounding is used, although other techniques have also been suggested (e.g., Johnston 1986).

The LP approach was developed to incorporate factors such as multiple stock lengths, a limit on the number of items in a pattern and ranges of demands instead of a single demand. This approach proved generally very useful for the so-called "easy" problems where there are many orders of small sizes. As there are many alternative solutions, the non-integer solutions are often close to the integer solutions. However, when there are few orders of a relatively large size (the so-called "hard" problems) the LP approach has not been very successful.

The 1-dimensional CSP can be viewed as a Bin Packing Problem(BPP). Given a set of items and a set of bins of fixed capacity, the BPP is to assign the items to the bins in such a manner that each item is assigned to one and only one bin and the total number of bins used is a minimum. Treating the bin capacity as stock length and the item sizes as order lengths the two problems are equivalent. BPP is NP-Complete. A number of good heuristics exist for solving BPP (Coffman et al 1984). Among the available exact algorithms for BPP, the branch and bound method of Martello & Toth (1990), using the reduction procedure is quite efficient. A good source of reference for all cutting and packing problems is Sweeney & Paternoster (1992).

Traditional methods of Operations Research such as integer programming and branch and bound have been used for the last few decades to obtain exact and approximate solutions to the cutting and packing problems. In the last few years, attention has been given to a number of innovative heuristic search techniques for these problems; these include Genetic Algorithms, Tabu Search and Simulated Annealing (Glover & Greenberg 1989; Falkenauer & Delchambre 1992; Hinterding & Juliff 1993; Prosser 1988; Reeves 1993; Smith 1985).

Contiguity. By solving a CSP by an LP model or otherwise, a set of patterns and the number of runs for each of these patterns are determined. To implement this solution at production floor level, the patterns have to be sequenced and scheduled. This sequencing factor is important for several reasons: the knife-setting changes required; the consistency in the quality of the products of the same customer order; and the storage of partly-finished and unready-for-packaging product stacks. Knife-setting difficulties can be ignored because automatic knife-changing devices are in use these days, but other problems of product quality and inventory are important. To alleviate these problems, it is desired that the production of items be, as far as practicable, "contiguous", ie., a particular size of item should be produced in successive patterns until that order size is completed. Similar requirements for other products like steel, glass, films, etc are quite possible.

Contiguous sequencing of patterns has not been well-researched. In practice, the sequencing is done in the 2nd step of a two-step process where in the 1st step, the patterns and their runs are determined. One example of 2-dimensional sequencing in this manner is (Yuen 1991).

One approach to deal with contiguity requirements is to include a measure of contiguity in the objective function of the cutting stock problem. For instance, the maximum number of partly-finished orders at any instant of a production run can serve as such a measure – the lower the number, the more contiguous the solution is. Adding a term "number of partly-finished open orders" into the objective function of a linear programming problem will make it non-linear and unwieldy for the existing LP methods for CSP. In this paper, the contiguity factor has been taken into account in some instances and solved by genetic algorithms.

3 Representing the Problem

Two different representations were used to solve the cutting stock problems using GAs. A Group based GA which uses a direct representation and an Order based GA which uses and indirect representation. The Group based GA uses a mapping which focuses on the groups in a solution. That is, it tries to find the best selection of the possible groups. A group is a selection of items which will be cut from a single stock length. The Order based GA focuses on the order of the items such that the items can be grouped into a solution using a decoder.

3.1 Mapping using a Group based GA

Falkenauer (1991) developed the Group based GA for the bin packing and other grouping problems. Here the emphasis is changed from the traditional GA where genes represent a single value and its position or order in relation to other genes is significant, to the situation where genes represent a group of items, and neither the order nor the position of the genes in the chromosome or the items in the gene is significant. This is a significant departure from the traditional GAs.

The problem in extending the Falkenauer's Bin Packing GA to the CSP where there are multiple stock lengths is how to encode from which stock length a group of items is to be cut from. Hinterding and Juliff used a multiple chromosome GA where a second chromosome was used to encode from which stock length the groups were to be cut. The second chromosome turns out to be unnecessary, as a valid group of items implies the stock length it should be cut from. This is the smallest stock length from which the group of items can successfully be cut.

Hence using this mapping we do not need to use a multi-chromosome GA for the Cutting Stock Problem.

In the Group based GA, each chromosome represents a number of groups of items such that all items to be cut are represented. Each gene represents a group of items, rather than a single item. Each group will be cut from a single stock length. This mapping is illustrated in Fig 1. The characteristics of this representation are that the number of genes is variable, the order of the genes in the chromosome has no significance, and the order of the items in each gene has no significance. These characteristics are compatible with the characteristics of the bin packing and cutting stock problems.

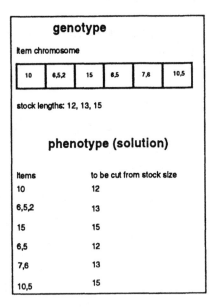

Fig. 1. Symbolic representation of mapping

In the Cutting Stock Problem with contiguity, the order of the genes becomes significant. The mapping does not need to change, but different crossover and mutation operators are used to reflect the significance of the order of the genes.

The encoder.

An intelligent encoder is used to build the initial population, and to build new groups for crossover and mutation operators. This encoder takes a list of stock lengths and a list of items in random order. Using a first fit algorithm it groups the items in the item list into groups to form genes. To build a group the encoder choses a stock length at random and then choses items without replacement from the list of items using a first fit algorithm. The stock length is not recorded in the chromosome, but is implied by the group itself (see above).

The reproduction operators.

Crossover. The crossover operator is a modification of Falkenauer's Grouping crossover (BPCX) (Falkenauer & Delchambre 1992). This crossover (called the Grouping crossover GCX) will work with chromosomes of different lengths, and does not depend on any ordering of the genes. It was designed so that the child can inherit meaningful information from both its parents. In this case it is the selection of genes (groups) the parents have.

The grouping crossover works in the following way: we randomly choose an insertion point in parent1 and a segment in parent2. The child is constructed by first copying into it the genes from parent1 up to the insertion point. Then we copy the genes from the segment in parent2 into the child, and lastly we copy the genes from parent1 after the insertion point into the child. We cannot blindly copy the genes into the child chromosome as a chromosome with duplicated items may result. A gene is only added to the child chromosome if all its items can be successfully subtracted from a list of items not yet included in the chromosome. At the end of crossover the list of items not yet included in the chromosome may not be empty, in this case the encoder is used to generate new genes (groups) from these items and the resulting gene(s) are added to the child chromosome. Therefore, if two identical parents are chosen for crossover, an identical child may not always result.

The Grouping crossover gives no significance to the order of the genes in the parents, so a new crossover was developed to give greater significance to the order of the genes for the CSP with contiguity. This new crossover, the Uniform Grouping crossover (UGCX) works as follows: we generate a template of randomly generated binary bits which has the same length as the first parent. We then copy for each position in parent1 the gene from that position into the child chromosome, and then the gene from that position from parent2 is also copied into the child if its corresponding value in the template is a 1 only. Again copying of a gene into the child chromosome is only carried out if the items from that genes can be successfully subtracted from a list of items not yet included in the child chromosome.

Mutation. The mutation operator is based on Falkenauer's group mutation operator (Falkenauer & Delchambre 1992). A number of genes are chosen and deleted. The items from the deleted genes are then used by the encoder to build new groups. These new groups are then added to the chromosome. The purpose of mutation in this case is to bring new groups into the chromosome. The genes to be deleted are chosen from those which do not cut exactly from the stock length (ie those with some wastage), and then randomly if there are insufficient of these. Note that we must delete at least two genes, as deleting only one gene and then rebuilding the chromosome would result in exactly the same chromosome. Our group mutation operator is different from Falkenauer's as he deleted the gene with the greatest wastage and then some others chosen at random. He adds the newly built genes to the end of the chromosome, while we insert them into the chromosome at a randomly chosen site.

For the cutting stock problem with contiguity we use our group mutation operator fifty percent of the time and we use the Remove and Reinsert mutation operator (RAR) for the other fifty percent of the time. The mutation operator is chosen randomly.

3.2 Mapping using an Order based GA

With this mapping the chromosome represents an ordering of all the items to be cut. A decoder is needed to construct the groups from the ordering of the items, and the GA will (hopefully) explore orderings which build the best groups.

A problem with this mapping is that we cannot represent the stock length the group is to be cut from unless we use a second chromosome. As we want to explore the effect of different mappings, we have restricted ourselves to not using a multi-chromosome GA. For this reason we shall use only single stock length problems for the order based GA.

The decoder. The choice of decoder is very important. If we use the next fit algorithm to build the groups, then crossover is too disruptive of the groups. The next fit algorithm is to take the next item from the chromosome and put it into the current group if it will fit and otherwise start a new group using the item. The basis of this problem is that a minor reordering of the items can lead to a major change in the decoded groups (Hinterding 1994), (Falkenauer 1991). The use of a first fit algorithm was successfully used to overcome this problem with the knapsack problem (Hinterding 1994) and it is used here as well.

The reproduction operators.

Crossover. We use the Uniform Order Based crossover (UOB) (Davis 1991), as relative rather than absolute order should be more significant.

Mutation. We use both swap and RAR mutation.

4　The Fitness Function

The evaluation function for the cutting stock problem is calculated as the result of the following cost function subtracted from the fitness ceiling of 1.0.

$$\text{cost} = \frac{1}{n} \left(\sum_{1,n} \sqrt{\frac{\text{waste}_i}{\text{stock\_length}_i}} + \frac{\text{number\_wasted}}{n} \right)$$

where

n =number of groups
stock_length$_i$ = the stock length that group$_i$ will be cut from
waste$_i$ = stock_length$_i$ − sum of items in group$_i$
number_wasted=number of stock lengths with wastage

The fitness function contains two terms. The first is to reduce the wastage, we take the square root of this term to give values near the limits of the range (0-1) extra weight. This is done as we wish to concentrate the wastage. The second term encourages solutions where fewer stock lengths contain wastage, as this leads to better solutions.

4.1　Stock cutting with contiguity

The evaluation function for the stock cutting problem with contiguity is calculated as the result of the following cost function subtracted from the fitness ceiling of 1.0.

$$\text{cost} = \frac{1}{10+n} \left(\sum_{1,n} \sqrt{\frac{\text{waste}_i}{\text{stock\_length}_i}} + 10 \left(\frac{\text{no\_open\_items}}{\text{no\_different\_item\_lengths}} \right)^2 \right)$$

where

n =number of groups
stock_length$_i$ = the stock length that group$_i$ will be cut from
waste$_i$ = stock_length$_i$ − sum of items in group$_i$

The second term in the fitness function is used to maximise the contiguity by minimising the number of open items. In fact we want to minimise the maximum number of open items. We square this term, to give values in the middle of the range (0-1) extra weight. The weight of 10 for the second term, was determined experimentally to give best results for Problem 4a.

As stated earlier, contiguity is difficult to quantify. By definition, a set of patterns is either contiguous or not. When an item once started is included in all subsequent runs until finished, the patterns are said to be sequenced contiguously. Ideally, any breach of this condition will destroy contiguity.

In practice however, perfect contiguity will not be the ultimate goal because it may raise the amount of wastage to an unacceptable level. Therefore, the opposing objective functions of waste minimisation and contiguity have been combined in the fitness function. The number of open items indicates the number of item sizes started to be produced, but not yet finished at the end of the current run. Fewer open items means better contiguity.

5 The GA

The basic Genetic Algorithm used in both cases is a steady-state GA based on the description of OOGA in Davis (1991). Tournament selection is used with a tournament size of 2 as this was faster and gave comparable results to roulette wheel selection with linear normalisation. It was developed using Smalltalk/V for Windows. The following parameters can be set:

- Population Size - set the size of the population.
- Duplicates - set a flag to allow or disallow duplicates to exist in the population. If duplicates are not allowed, any duplicates produced by reproduction are discarded while they still count as an evaluation. We determine whether two chromosomes are the same by comparing their genotypes.
- Number of Evaluations - set the number of evaluations for the run. We use evaluations rather than generations so that we can compare between runs where the population size and replacement rate are different.
- Replacement Rate - set the percentage of the population that will be replaced by reproduction in one generation. The rate can be set from 0 to 100%.
- Crossover Rate - set the percentage of the replacement population that will be replaced by crossover in one generation. The remainder of the replacement population will be produced by mutation. The rate can be set from 0 to 100%.
- Poisson Mutation - use a Poisson distributed random variable to determine how many genes to mutate in a chromosome.
- Poisson Mean - set the mean (λ) for the Poisson distributed random variable.

In the Genetic Algorithm used, a new chromosome is produced either by crossover or mutation but not both. This was done so that the separate effects of these reproduction operators could be determined.

The mutation rate for the GA is 100 - Crossover rate. The mutation rate is the percentage of chromosomes of the replacement population that will undergo mutation. If Poisson Mutation is false, then mutation of one gene is generally carried out. The exceptions are swap mutation where two genes are swapped, and grouping mutation where three genes will be deleted and rebuilt. If Poisson Mutation is true, then the number of genes to be mutated in a chromosome is determined by sampling a Poisson distributed random variable with mean λ.

6 Results

For this study we used mainly five different problems (App. A), the first three (Problems 1-3) are from Hinterding & Juliff (1993). Problem 4 was derived from

Problem 3 and modified to make it harder. Problem 5 was taken from a paper by Goulimis (1990). Two versions of each problem were created. One version has multiple stock lengths, and the other uses only a single stock length. The single stock length problems have an "a" suffix.

All the results were produced by averaging the results of 20 runs. Each run is divided into an equal number of intervals. The mean and standard deviation of the fitness values are calculated for each of the intervals over the batch of 20 runs. The parameters for each of the variants of the GA was optimised for best performance on Problem 4a, and run on all the other problems using the same settings. Typical solutions are shown in App. B.

We have dramatically improved on the results for Problems 1-3 as reported by Hinterding & Juliff. They were unable to solve these problems completely within 20,000 evaluations, our results show these problems solved completely in well under 1,000 evaluations.

We note that for the cutting stock problem without considering contiguity the Group based GA is clearly better than the Order based GA. In Fig. 2, we graph the percentage crossover used against the maximum fitness reached for Problem 4a using the order based GA. This graph illustrates that crossover degrades the performance of the GA.

Problem 4a

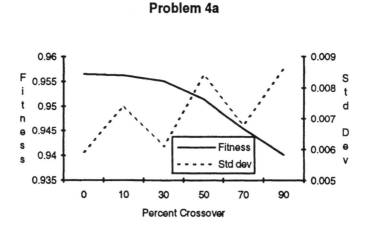

Fig. 2. Order Based GA, percentage crossover vs average fitness

For the cutting stock problem with contiguity, both GAs give comparable results while the Group based GA is slightly better. To achieve good results, the number of evaluations had to be increased significantly.

Falkenauer (1994) improved his Bin Packing GA by using local optimisation inspired by the reduction method of Martello & Toth (1990). His paper gives details of generating very difficult bin packing problems. We ran the Group based GA on a number of these difficult problems. For each of these problems the optimum solution is known. On the 30 item problems we were able to solve it completely 75% of the time, and on 60 item problems 25% of the time. With 100 item problems we were able to get within one stock length of the optimum within 5,000 evaluations. We conclude that his local optimisation method is extremely effective as Falkenauer was able to solve completely problems of up to 501 items. We also note that this local optimisation method cannot be used with the Order based GA, as the groups do not exist in the chromosome.

Reeves (1993) also used a reduction procedure for solving the bin packing problem by a GA. He reduced the problem size by deleting that subset of items which could fill one or more bins completely (or near completely). This type of reduction can be done with Order based mappings, and is easy to execute and can save substantial computation time. However, as Reeves acknowledges, it can lead to suboptimal solutions. In fact, the greater the number of items deleted by this reduction procedure, the greater the risk and level of suboptimality. Falkenauer's use of Martello & Toth's reduction does not lead to suboptimality.

Table 1. Results: Group based, no contiguity

Group based, no contiguity
Population: 75 Allow duplicates: false Replacement rate: 50%
Crossover rate: 30% Poisson mutation: ($\lambda = 4$)
Crossover: grouping
Mutation: grouping

Problem	Evaluations	Mean fitness	Std dev	Found at
1	1184	1.0	0.0	407
2	1184	1.0	0.0	740
3	1184	1.0	0.0	407
4	2294	0.9995	0.0022	2294
5	2294	0.9998	0.0007	2294

Problem	Evaluations	Mean fitness	Std dev	Found at
1a	1184	0.9133	0.0	296
2a	1184	0.9227	0.0018	1184
3a	1184	1.0	0.0	407
4a	1184	0.9642	0.0	851
5a	2294	0.8479	0.007	2294

7 Discussion

We can see from the results for the cutting stock problem without contiguity that the Group based GA is significantly better than the Order based GA. It

Table 2. Results: Order based, no contiguity

Order based, no contiguity
Population: 100 Allow duplicates: false Replacement rate: 10%
Crossover rate: 0% Poisson mutation: ($\lambda = 4$)
Crossover: UOB
Mutation: RAR

Problem	Evaluations	Mean fitness	Std dev	Found at
1a	1200	0.9133	0.0	320
2a	3200	0.9198	0.0029	3200
3a	4200	1.0	0.0	2560
4a	5200	0.9588	0.0063	5200
5a	1020	0 0.8489	0.0048	10200

is also evident that with the Order based GA the use of crossover degrades the solutions found. Other order based crossovers have been tried but better results were obtained with the Uniform Order based crossover. We attribute this result to the fact that while there are orderings which will map to the best solution, this is a many to one mapping and there is no ordering information which the crossover operator can exploit. Any permutation of the groups in a solution is an equivalent solution, as is any permutation of the items in any group.

This is in contrast to the Knapsack Problem results in Hinterding (1994) where the Order based GA gave better results than the Selection based GA. Here the ordering performed was to move "better" genes towards the front of the chromosome.

In fact, the results of the Order based GA on the CSP could be interpreted as indicating that reordering performed by the crossover operator interfered with the mutation operator. This suggests that there is no ordering information to process using this mapping.

For the cutting stock problem with contiguity the relative order of the groups becomes significant. The performance of the Order based and Group based GAs is roughly the same on this problem, with the Group based GA giving slightly better results. The Order based GA now has ordering information to process, as best results were obtained using 90% crossover. The crossover operator for the Group based based GA was modified so that ordering information could be processed. This modification was successful as best results were obtained with a crossover rate of 70%.

Comparing the results for the CSP with contiguity to the CSP without contiguity, we see they differ in three ways for a CSP with contiguity:

- groups with the same pattern are brought together.
- the number of distinct item lengths in groups are reduced.
- the number of distinct patterns are reduced.

Table 3. Results: Group based, with contiguity.

Group based, with contiguity
Population: 100 Allow duplicates: false Replacement rate: 70%
Crossover rate: 70% Poisson mutation: ($\lambda = 4$)
Crossover: UGCX
Mutation: 50% grouping, 50% RAR

Problem	Evaluations	Mean fitness	Std dev	Found at
1	2155	0.9860	0.0031	2155
2	5275	0.9796	0.013	5275
3	4235	0.9828	0.0041	4235
4	5275	0.9610	0.014	5275
5	10435	0.9889	0.0073	10435

Problem	Evaluations	Mean fitness	Std dev	Found at
1a	2155	0.956	0.0	1739
2a	5275	0.9212	0.0171	5275
3a	4235	0.9387	0.0318	4235
4a	5275	0.9335	0.0103	5275
5a	10600	0.8469	0.0118	10600

Table 4. Results: Order based, with contiguity.

Order based, with contiguity
Population: 100 Allow duplicates: false Replacement rate: 70%
Crossover rate: 90% Poisson mutation: ($\lambda = 5$)
Crossover: UOB
Mutation: RAR

Problem	Evaluations	Mean fitness	Std dev	Found at
1a	2200	0.9554	0.0021	1990
2a	5700	0.9190	0.0109	5140
3a	4300	0.9096	0.0127	4300
4a	5700	0.9236	0.0145	5700
5a	10600	0.8454	0.0132	10600

All these features are desirable for the solution, but only the first can be achieved by the second operation of the traditional two-step process.

From the results for the CSP with contiguity, we see that there are some problems with the fitness function. There are solutions with the same fitness value, but differing contiguity (see App. B). This occurs because we are optimising two conflicting measures, wastage and contiguity. If we did not worry about wastage, the contiguity problem is trivial. The solution is: cut the items in item length order ignoring wastage; perfect contiguity results. The conflict occurs as to minimise wastage, the GA attempts to concentrate wastage, while to maximise contiguity the wastage may have to be distributed. These conflicting aims can affect the quality of the solutions.

8 Conclusion

We have shown that the Cutting Stock Problem with and without contiguity can be successfully solved using Genetic Algorithms. It was relatively easy to modify the Group based and Order based GAs to include consideration of contiguity. Traditional OR methods do not handle CSP with contiguity well.

We have shown the results of different mappings and crossovers for GAs on these problems. The Group based GA is better for the Cutting Stock Problem. For the Cutting Stock Problem with contiguity both the GAs work equally well. Only the Group based GA can be modified to include the local optimisation used by Falkenauer.

A pattern now starts to emerge as to the suitability of various mappings for different grouping problems. From Hinterding (1994) we see that an Order based GA is superior as the Knapsack Problem can be successfully mapped into an ordering problem. For CSP without contiguity the Group based GA is superior to the Order based GA, as it contains no ordering information for it to process. Lastly for CSP with contiguity both GAs are equally successful: the Order based GA was successful it has ordering information to process; the Group based GA was successful because the crossover was modified so that it could process ordering information.

References

Coffman, E. G., Garey, M. R. and Johnson, D. S.: Approximation Algorithms for Bin-Packing - an Updated Survey, in G. Ausiello, M. Lucertini and P. Serafini (eds), Algorithm Design for Computer System Design, Springer Verlag, Vienna(1984), pp.49-106.

Davis, L.(ed): Handbook of Genetic Algorithms, Van Nostrand Reinhold 1991.

Dyckhoff, H.: A Typology of Cutting and Packing Problems, European Journal of Operational Research, Vol 44 (1990), pp. 145-159.

Falkenauer, E,: A Genetic Algorithm for Grouping, Proceedings of the Fifth International Symposium on Applied Stochastic Models and Data Analysis, Granada, Spain 1991.

Falkenauer, E., & Delchambre, A.: A Genetic Algorithm for Bin Packing and Line Balancing, Proceedings of 1992 IEEE International Conference on Robotics and Automation(RA92), pp. 1186-1193, Nice 1992.

Falkenauer, E.: Setting New Limits in Bin Packing with a Grouping GA Using Reduction, Technical Report RO108, Department of Industrial Automation, Research Centre for the Belgium Metalworking Industry, Brussels Belgium, 1994.

Gilmore, P. C. and Gomory, R. E.: A Linear Programming Approach to the Cutting Stock Problem; Part II, Operations Research, Vol 11 (1963), pp 863-888.

Glover, F. and Greenberg, H. J.: New Approaches for Heuristic Search: A Bilateral Linkage with Artificial Intelligence, Vol 39 (1989), pp.119-130.

Goldberg, D. E. and Lingle, R.: Alleles, Loci, and the Travelling Salesman Problem. Proceedings of an International Conference on Genetic Algorithms and their Applications pp 154-159, 1985.

Goldberg, D. E.: Genetic Algorithms in Search, Optimization & Machine Learning, Addison-Wesley 1989.

Goulimis, C.: Optimal solutions for the cutting stock problem, European Journal of Operational Research, Vol 44 (1990), pp. 197-208.

Hinterding, R. H., and Juliff, K. C.: A Genetic Algorithm for Stock Cutting: An exploration of Mapping Schemes, Technical Report 24COMP3, Department of Computer and Mathematical Sciences, Victoria University of Technology, Victoria Australia, Feb. 1993.

Hinterding, R. H.: Mapping, Order-independent Genes and the Knapsack Problem, Proceedings of the First IEEE Conference on Evolutionary Computation (ICEC'94), pp. 13-17, Orlando 1994.

Johnston, R.E.: Rounding Algorithms for Cutting Stock Problems, Asia-Pacific Journal of Operational Research, Vol 3 (1986), pp 166-171.

Juliff, K.: A multi-chromosome genetic algorithm for pallet loading, Proceedings of the Fifth International Conference on Genetic Algorithms, pp 476-73, Urbana-Champaign 1993.

Prosser, P.: A Hybrid Genetic Algorithm for Pallet Loading, Proceedings of 8th European Conference on Artificial Intelligence, London 1988.

Martello, S. and Toth, P.: Knapsack Problems, John Wiley & Sons 1990.

Reeves, C.: Hybrid Genetic Algorithms for Bin-packing and Related Problems, submitted to: Annals of Operations Research 1993.

Smith, D.: Bin Packing with adaptive search, Proceedings of an International Conference on Genetic Algorithms pp.202-206, 1985.

Syswerda, G.: Schedule Optimization Using Genetic Algorithms, Handbook of Genetic Algorithms, Davis, L. (ed), 1991, Van Nostrand Reinhold.

Sweeney, P. E. & Paternoster, E. R.: A Categorized, Application-Oriented Research Bibliography on Cutting and Packing Problems, Journal of the Operational Research Society, Vol 43 (1992), pp.691-706.

Yuen, B. J.: Heuristics for Sequencing Cutting Patterns, European Journal of Operational Research, Vol 55 (1991), pp 183-190.

Appendix

A The Problems

Problem 1: stock lengths 10, 13, 15
Problem 1a: stock length 14
20 items

Item Length	3	4	5	6	7	8	9	10
No. req.	5	2	1	2	4	2	1	3

Problem 2: stock lengths 10, 13, 15
Problem 2a: stock length 15
50 items

Item Length	3	4	5	6	7	8	9	10
No. req.	4	8	5	7	8	5	5	8

Problem 3: stock lengths 10, 13, 15, 20, 22, 25
Problem 3a: stock length 25
60 items

Item Length	3	4	5	6	7	8	9	10
No. req.	6	12	6	5	15	6	4	6

Problem 4: stock lengths 13, 20, 25
Problem 4a: stock length 25
60 items

Item Length	5	6	7	8	9	10	11	12
No. req.	7	12	15	7	4	6	8	1

Problem 5: stock lengths 4300, 4250, 4150, 3950, 3800, 3700, 3550, 3500
Problem 5a: stock length 4300
126 items

Item Length	2350	2250	2220	2100	2050	2000	1950	1900	1850
No. req.	2	4	4	15	6	11	6	15	13
Item Length	1700	1650	1350	1300	1250	1200	1150	1100	1050
No. req.	5	2	9	3	6	10	4	8	3

B Typical Solutions

Problem 1 without contiguity.
Ord: 7 Ord: 8 Wastage: 0 | 15
Ord: 3 Ord: 10 Wastage: 0 | 13
Ord: 3 Ord: 10 Wastage: 0 | 13
Ord: 7 Ord: 8 Wastage: 0 | 15
Ord: 6 Ord: 7 Wastage: 0 | 13
Ord: 3 Ord: 10 Wastage: 0 | 13
Ord: 4 Ord: 5 Ord: 6 Wastage: 0 | 15
Ord: 4 Ord: 9 Wastage: 0 | 13
Ord: 3 Ord: 3 Ord: 7 Wastage: 0 | 13
Total Wastage: 0 No. waste items: 0 Stock used: 9

Problem 1 with contiguity
Ord: : 4 : 9 Wastage: 0 | 13 Open: 1
Ord: : 3 : 3 : 3 : 4 Wastage: 0 | 13 Open: 1
Ord: : 3 : 3 : 7 Wastage: 0 | 13 Open: 1
Ord: : 6 : 7 Wastage: 0 | 13 Open: 2
Ord: : 6 : 7 Wastage: 0 | 13 Open: 1
Ord: : 7 : 8 Wastage: 0 | 15 Open: 1
Ord: : 5 : 8 Wastage: 0 | 13 Open: 0
Ord: : 10 Wastage: 0 | 10 Open: 1
Ord: : 10 Wastage: 0 | 10 Open: 1
Ord: : 10 Wastage: 0 | 10 Open: 0
Total Wastage: 0 No. waste items: 0 Stock used: 10 No. different orders: 8

Problem 1a without contiguity
Ord: : 3 : 3 : 8 Wastage: 0 | 14 Open: 2
Ord: : 5 : 9 Wastage: 0 | 14 Open: 2
Ord: : 4 : 10 Wastage: 0 | 14 Open: 4
Ord: : 7 : 7 Wastage: 0 | 14 Open: 5
Ord: : 3 : 3 : 8 Wastage: 0 | 14 Open: 4
Ord: : 7 : 7 Wastage: 0 | 14 Open: 3
Ord: : 4 : 10 Wastage: 0 | 14 Open: 2
Ord: : 6 : 6 Wastage: 2 | 14 Open: 2
Ord: : 3 : 10 Wastage: 1 | 14 Open: 0
Total Wastage: 3 No. waste items: 2 Stock used: 9 No. different orders: 8

Problem 1a with contiguity
Ord: : 4 : 10 Wastage: 0 | 14 Open: 2
Ord: : 4 : 10 Wastage: 0 | 14 Open: 1
Ord: : 3 : 10 Wastage: 1 | 14 Open: 1
Ord: : 3 : 3 : 8 Wastage: 0 | 14 Open: 2
Ord: : 3 : 3 : 8 Wastage: 0 | 14 Open: 0
Ord: : 6 : 6 Wastage: 2 | 14 Open: 0
Ord: : 7 : 7 Wastage: 0 | 14 Open: 1
Ord: : 7 : 7 Wastage: 0 | 14 Open: 0
Ord: : 5 : 9 Wastage: 0 | 14 Open: 0
Total Wastage: 3 No. waste items: 2 Stock used: 9 No. different orders: 8

Problem 4 without contiguity
Ord: 6 Ord: 7 Ord: 7 Wastage: 0 | 20
Ord: 10 Ord: 10 Wastage: 0 | 20
Ord: 9 Ord: 11 Wastage: 0 | 20
Ord: 5 Ord: 9 Ord: 11 Wastage: 0 | 25
Ord: 8 Ord: 12 Wastage: 0 | 20
Ord: 9 Ord: 11 Wastage: 0 | 20
Ord: 6 Ord: 7 Wastage: 0 | 13
Ord: 5 Ord: 8 Wastage: 0 | 13
Ord: 5 Ord: 8 Wastage: 0 | 13
Ord: 6 Ord: 7 Wastage: 0 | 13
Ord: 6 Ord: 8 Ord: 11 Wastage: 0 | 25
Ord: 6 Ord: 7 Wastage: 0 | 13
Ord: 7 Ord: 7 Ord: 11 Wastage: 0 | 25
Ord: 6 Ord: 7 Wastage: 0 | 13
Ord: 5 Ord: 9 Ord: 11 Wastage: 0 | 25
Ord: 5 Ord: 8 Wastage: 0 | 13
Ord: 6 Ord: 7 Wastage: 0 | 13
Ord: 6 Ord: 7 Wastage: 0 | 13
Ord: 6 Ord: 8 Ord: 11 Wastage: 0 | 25
Ord: 5 Ord: 10 Ord: 10 Wastage: 0 | 25
Ord: 6 Ord: 7 Wastage: 0 | 13
Ord: 6 Ord: 7 Wastage: 0 | 13
Ord: 5 Ord: 8 Wastage: 0 | 13
Ord: 6 Ord: 7 Wastage: 0 | 13
Ord: 10 Ord: 10 Wastage: 0 | 20
Ord: 7 Ord: 7 Ord: 11 Wastage: 0 | 25
Total Wastage: 0 No. waste items: 0 Stock used: 26

Problem 4 with contiguity (Solution A)
Generation: 100 Best at: 79 Ave fit: 0.954 Max fit: 0.972 Min fit: 0.878
REACHED MAX GENERATION Best at: 79
Ord: : 5 : 8 Wastage: 0 | 13 Open: 2
Ord: : 5 : 8 Wastage: 0 | 13 Open: 2
Ord: : 5 : 8 Wastage: 0 | 13 Open: 2
Ord: : 5 : 8 Wastage: 0 | 13 Open: 2
Ord: : 5 : 8 Wastage: 0 | 13 Open: 2
Ord: : 5 : 8 Wastage: 0 | 13 Open: 2
Ord: : 5 : 8 Wastage: 0 | 13 Open: 0
Ord: : 10 : 10 Wastage: 0 | 20 Open: 1
Ord: : 10 : 10 Wastage: 0 | 20 Open: 1
Ord: : 10 : 10 Wastage: 0 | 20 Open: 0
Ord: : 6 : 7 Wastage: 0 | 13 Open: 2
Ord: : 6 : 7 Wastage: 0 | 13 Open: 2
Ord: : 6 : 7 : 7 Wastage: 0 | 20 Open: 2
Ord: : 6 : 7 Wastage: 0 | 13 Open: 2
Ord: : 6 : 7 Wastage: 0 | 13 Open: 2
Ord: : 6 : 7 Wastage: 0 | 13 Open: 2
Ord: : 6 : 7 Wastage: 0 | 13 Open: 2
Ord: : 6 : 7 Wastage: 0 | 13 Open: 2
Ord: : 6 : 7 Wastage: 0 | 13 Open: 2
Ord: : 6 : 7 Wastage: 0 | 13 Open: 2
Ord: : 6 : 7 Wastage: 0 | 13 Open: 2
Ord: : 6 : 7 Wastage: 0 | 13 Open: 1
Ord: : 7 : 7 : 11 Wastage: 0 | 25 Open: 1
Ord: : 11 : 12 Wastage: 2 | 25 Open: 1
Ord: : 11 : 11 Wastage: 3 | 25 Open: 1
Ord: : 9 : 11 Wastage: 0 | 20 Open: 2
Ord: : 9 : 11 Wastage: 0 | 20 Open: 2
Ord: : 9 : 11 Wastage: 0 | 20 Open: 2
Ord: : 9 : 11 Wastage: 0 | 20 Open: 0
Total Wastage: 5 No. waste items: 2 Stock used: 29 No. different orders: 8

Problem 4 with contiguity (Solution B)
Generation: 100 Best at: 68 Ave fit: 0.944 Max fit: 0.972 Min fit: 0.856
REACHED MAX GENERATION Best at: 68
Ord: : 5 : 7 : 8 Wastage: 0 | 20 Open: 3
Ord: : 5 : 7 : 8 Wastage: 0 | 20 Open: 3
Ord: : 5 : 7 : 8 Wastage: 0 | 20 Open: 3
Ord: : 5 : 6 : 6 : 8 Wastage: 0 | 25 Open: 4
Ord: : 5 : 6 : 6 : 8 Wastage: 0 | 25 Open: 4
Ord: : 5 : 6 : 6 : 8 Wastage: 0 | 25 Open: 4
Ord: : 5 : 6 : 6 : 8 Wastage: 0 | 25 Open: 2
Ord: : 6 : 7 Wastage: 0 | 13 Open: 2
Ord: : 6 : 7 : 12 Wastage: 0 | 25 Open: 2
Ord: : 6 : 7 Wastage: 0 | 13 Open: 2
Ord: : 6 : 7 Wastage: 0 | 13 Open: 1
Ord: : 10 : 10 Wastage: 0 | 20 Open: 2
Ord: : 10 : 10 Wastage: 0 | 20 Open: 2
Ord: : 10 : 10 Wastage: 0 | 20 Open: 1
Ord: : 7 : 7 : 11 Wastage: 0 | 25 Open: 2
Ord: : 7 : 7 : 11 Wastage: 0 | 25 Open: 2
Ord: : 7 : 7 : 11 Wastage: 0 | 25 Open: 2
Ord: : 7 : 7 : 11 Wastage: 0 | 25 Open: 1
Ord: : 9 : 11 Wastage: 0 | 20 Open: 2
Ord: : 9 : 11 Wastage: 0 | 20 Open: 2
Ord: : 9 : 11 Wastage: 0 | 20 Open: 2
Ord: : 9 : 11 Wastage: 0 | 20 Open: 0
Total Wastage: 0 No. waste items: 0 Stock used: 22 No. different orders: 8

Note: A comparison of the two solutions (A & B) for the same problem illustrates the effect of contiguity. While both solutions have the same fitness (0.972), A has a wastage of 5 and B has 0 - clearly B is better than A in terms of wastage. However, A has a maximum of 2 items open, compared to 4 of B; thus A is superior in terms of contiguity. This occurs due to the conflicting criteria of wastage and contiguity. It should also be noted that the different numbers of stock used (29 & 22) has little significance as differing stock lengths are used in the solutions.

Problem 4a without contiguity
Ord: : 5 : 9 : 11 Wastage: 0 | 25 Open: 3
Ord: : 7 : 8 : 10 Wastage: 0 | 25 Open: 6
Ord: : 5 : 10 : 10 Wastage: 0 | 25 Open: 6
Ord: : 5 : 6 : 7 : 7 Wastage: 0 | 25 Open: 7
Ord: : 6 : 8 : 11 Wastage: 0 | 25 Open: 7
Ord: : 5 : 9 : 11 Wastage: 0 | 25 Open: 7
Ord: : 7 : 7 : 11 Wastage: 0 | 25 Open: 7
Ord: : 6 : 9 : 10 Wastage: 0 | 25 Open: 7
Ord: : 7 : 7 : 11 Wastage: 0 | 25 Open: 7
Ord: : 7 : 7 : 11 Wastage: 0 | 25 Open: 7
Ord: : 6 : 9 : 10 Wastage: 0 | 25 Open: 6
Ord: : 6 : 7 : 12 Wastage: 0 | 25 Open: 6
Ord: : 6 : 6 : 6 : 7 Wastage: 0 | 25 Open: 6
Ord: : 6 : 8 : 11 Wastage: 0 | 25 Open: 6
Ord: : 5 : 6 : 7 : 7 Wastage: 0 | 25 Open: 6
Ord: : 7 : 8 : 10 Wastage: 0 | 25 Open: 5
Ord: : 6 : 8 : 11 Wastage: 0 | 25 Open: 4
Ord: : 5 : 5 : 7 : 8 Wastage: 0 | 25 Open: 2
Ord: : 6 : 8 Wastage: 11 | 25 Open: 0
Total Wastage: 11 No. waste items: 1 Stock used: 19 No. different orders: 8

Problem 4a with contiguity
Ord: : 7 : 7 : 11 Wastage: 0 | 25 Open: 2
Ord: : 7 : 7 : 11 Wastage: 0 | 25 Open: 2
Ord: : 7 : 7 : 11 Wastage: 0 | 25 Open: 2
Ord: : 7 : 7 : 11 Wastage: 0 | 25 Open: 2
Ord: : 7 : 7 : 11 Wastage: 0 | 25 Open: 2
Ord: : 7 : 7 : 11 Wastage: 0 | 25 Open: 2
Ord: : 7 : 7 : 11 Wastage: 0 | 25 Open: 2
Ord: : 11 : 12 Wastage: 2 | 25 Open: 1
Ord: : 7 : 9 : 9 Wastage: 0 | 25 Open: 1
Ord: : 5 : 10 : 10 Wastage: 0 | 25 Open: 3
Ord: : 5 : 10 : 10 Wastage: 0 | 25 Open: 3
Ord: : 5 : 10 : 10 Wastage: 0 | 25 Open: 2
Ord: : 8 : 9 Wastage: 8 | 25 Open: 3
Ord: : 8 : 8 : 9 Wastage: 0 | 25 Open: 2
Ord: : 5 : 6 : 6 : 8 Wastage: 0 | 25 Open: 3
Ord: : 5 : 6 : 6 : 8 Wastage: 0 | 25 Open: 3
Ord: : 5 : 6 : 6 : 8 Wastage: 0 | 25 Open: 3
Ord: : 5 : 6 : 6 : 8 Wastage: 0 | 25 Open: 1
Ord: : 6 : 6 : 6 : 6 Wastage: 1 | 25 Open: 0
Total Wastage: 11 No. waste items: 3 Stock used: 19 No. different orders: 8

GASBOR: A Genetic Algorithm for Switchbox Routing in Integrated Circuits

Jens Lienig and K. Thulasiraman

Department of Electrical and Computer Engineering
Concordia University
1455 de Maisonneuve Blvd. West
Montreal, Quebec H3G 1M8, Canada
E-mail: jensl@ece.concordia.ca

Abstract. A new genetic algorithm for switchbox routing in the physical design process of integrated circuits is presented. Our algorithm, called GASBOR, is based on a three-dimensional representation of the switchbox and problem-specific genetic operators. The performance of the algorithm is tested on different benchmarks and it is shown that the results obtained using the proposed algorithm are either qualitatively similar to or better than the best published results.

1 Introduction

Routing is one of the major tasks in the physical design process of integrated circuits. Pins that belong to the same signal net are connected with each other subject to a set of routing constraints during the routing procedure.

A switchbox is a rectangular routing region of fixed size on an integrated circuit. A switchbox contains pins located on all four boundaries. An example of a simple switchbox with a possible routing solution is shown in Figure 1.

The problem of switchbox routing is twofold: (1) to determine if a valid routing exists within the boundaries of the switchbox, and if so, (2) to optimize the routing structure according to certain quality factors. Both problems are considered as NP-complete. Although different algorithms have been proposed over the years (e.g., [1]-[4], [7], [10], [11]), the problem of finding a globally optimized routing solution of a switchbox is still open.

Genetic algorithms (GAs) are a new class of heuristic search methods based on the biological evolution model. During the last few years, GAs have been applied more and more successfully to find good heuristic solutions to NP-complete optimization problems [6].

To our knowledge, only two papers have been published in which strategies derived from the concept of GAs are applied to switchbox routing [3],[10]. The router in [3] combines the so-called steepest descent method with features of GAs. The crossover operator, however, is restricted to the exchanging of entire nets and the mutation procedure performs only the creation of new initial individuals. In [10], a rip-up-and-rerouter that is based on a probabilistic rerouting of nets of one routing structure is presented. However, the routing is carried out

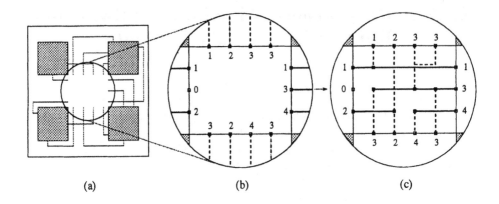

Fig. 1. An example of an integrated circuit (a) with a switchbox routing problem (b), and a possible routing solution of the switchbox (c). Dashed lines represent interconnections on the first layer, and solid lines represent interconnections on the second layer.

by a deterministic Lee algorithm [8]. Furthermore, main components of GAs, such as the crossover of different individuals, are not applied.

In this paper, we present a new GA for switchbox routing, called GASBOR, that is fundamentally different from the above mentioned approaches. First, our algorithm assumes that the switchbox is extendable in both directions. Subsequently, these extensions are reduced with the goal to reach the fixed size of the switchbox. Second, our algorithm is based on a problem-specific lattice-like representation of the switchbox instead of the commonly used binary string representation scheme in traditional GAs. Third, all routing structures are created randomly with a special random, rather than deterministic, routing strategy. And fourth, in contrast to [3], where the crossover operator exchanges only entire nets, our genetic operators work on the lowest level of a routing structure, namely, its grid points.

The contributions of this paper are:

- To formulate a GA that is capable of handling the routing problem of switchboxes.
- To compare the performance of our algorithm with previous approaches and describe the advantages and limitations of our strategy.
- To provide suggestions for other applications of GAs in the physical layout design of integrated circuits.

2 Problem Description

The switchbox routing problem is defined as follows. Consider a rectangular routing region, called *switchbox*, with a number of *pins* located on all four boundaries.

The pins that belong to the same net have to be connected subject to certain constraints and quality factors. The connection has to be made inside the boundaries of the switchbox on a symbolic routing area consisting of horizontal *rows* and vertical *columns*.

The interconnections are associated with the following constraints:

- Two layers are available for routing (see Figure 1).
- A net may change from one layer to the other one using a contact window called *via*.
- Different nets cannot cross each other on the same layer and must respect a minimum distance.
- The perimeter of the switchbox is not used for routing.

Three quality factors are used in this work to assess the quality of the routing result:

- Routing area
  The routing area, expressed as the number of rows and columns, is minimized during the evolutionary process until it reaches the fixed size of the switchbox. A final size larger than the fixed size means an unroutable solution.
- Net length
  The shorter the length of the interconnection nets the smaller the propagation delay.
- Number of vias
  The fewer the number of vias the better the routing quality.

3 Description of the Algorithm

3.1 Survey

GAs, in general, carry out optimization by simulating biological evolutionary processes. A population of individuals representing different problem solutions is subjected to genetic operators, such as, selection, crossover and mutation that are derived from the model of evolution. Using these operators the individuals are steadily improved over many generations and eventually the best individual resulting from this process is presented as the best solution to the problem.

An overview of the GA presented in this paper is shown in Figure 2. The number of individuals $|\mathcal{P}_c|$ is kept constant throughout all generations. Our mutation operator is applied after the reduction procedure, i.e., the modifications caused by the mutation operator remain "unpunished" in the population during the next mate selection and crossover procedure. This separation of the crossover and mutation procedures improves the ability of our approach to overcome local optima. Since the mutation operator has access to all individuals, the best individual is saved in each generation before the mutation operator is applied. At the end of the algorithm, the best individual p_{best} that has ever existed constitutes our final routing solution.

```
create initial population (𝒫c)
fitness_calculation (𝒫c)
pbest = best_individual (𝒫c)
for generation = 1 until max_generation
    𝒫n = ∅
    for offspring = 1 until max_descendant
        pα = selection (𝒫c)
        pβ = selection (𝒫c)
        𝒫n = 𝒫n ∪ crossover (pα, pβ)
    endfor
    fitness_calculation (𝒫n)
    𝒫c = reduction (𝒫c ∪ 𝒫n)
    pbest = best_individual (pbest ∪ 𝒫c)
    mutation (𝒫c)
    fitness_calculation (𝒫c)
endfor
routing solution = best_individual (pbest ∪ 𝒫c)
```

Fig. 2. Outline of the algorithm.

3.2 Creation of an Initial Population

The initial population is constructed from randomly created individuals.

First, each of these individuals is assigned a random initial row number y_{ind} and a random initial column number x_{ind}.

Let $\mathcal{S} = \{s_1, ...s_i, ...s_k\}$ be the set of all pins of the switchbox which are not connected yet and let $\mathcal{T} = \{t_1, ...t_j, ...t_l\}$ be the set of all pins having at least one connection to another pin. Initially $\mathcal{T} = \emptyset$. A pin $s_i \in \mathcal{S}$ is chosen randomly among all elements in $\mathcal{S}$. If $\mathcal{T}$ contains pins $\{t_u, ...t_j, ...t_v\}$ (with $1 \leq u < v \leq l$) of the same net, a pin t_j is randomly selected among them. Otherwise a second pin of the same net is randomly chosen from $\mathcal{S}$ and transferred into $\mathcal{T}$. Both pins (s_i, t_j) are connected with a so-called "random routing". Then s_i is transferred into $\mathcal{T}$. The process continues with the next random selection of $s_i \in \mathcal{S}$ until $\mathcal{S} = \emptyset$.

The random routing of two points (s_i, t_j) is based on our random routing strategy proposed in [9]. Its main characteristic is a line-search algorithm with random positions of alternate horizontal and vertical extension lines. A random routing of (s_i, t_j) is illustrated in Figure 3.

The extension is stopped when

- the extension lines of both points meet each other on the same layer, or
- the extension lines of s_i touch a net point which is already connected with t_j as shown in Figure 3 (e) (or vice versa).

191

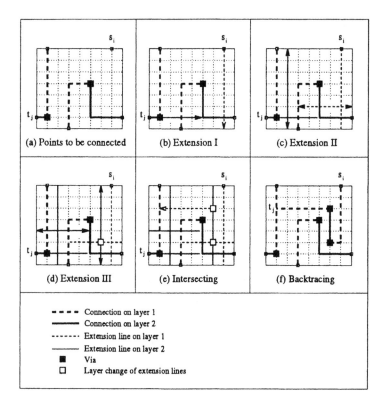

Fig. 3. Random routing of (s_i, t_j).

In the latter case, t_j (or s_i) is replaced with this meeting point (see Figure 3 (f)).

If the creation of extension lines does not succeed in one of these conditions within a certain number of iterations, all extension lines are erased and the switchbox is extended either with an additional row on a random position y_{add} with $1 \leq y_{add} \leq y_{ind}$ or an additional column on a random position x_{add} with $1 \leq x_{add} \leq x_{ind}$. After adjusting all previous routed nets to this new row and column, respectively, s_i and t_j undergo a new attempt to connect them with randomly created extension lines.

Tracing the shortest path on the extension lines between s_i and t_j (see Figure 3 (e,f)) avoids unnecessary loops in the connection without sacrificing the randomness of the resulting routing path.

The creation of the initial population is finished when the number of completely routed switchboxes is equal to the population size $|\mathcal{P}_c|$. As a consequence of our strategy, these initial individuals are quite different from each other and scattered all over the search space.

3.3 Calculation of Fitness

The fitness $F(p)$ of each individual $p \in \mathcal{P}$ is calculated to assess the quality of its routing structure relative to the rest of the population $\mathcal{P}$. The selection of the mates for crossover and the selection of individuals which are transferred into the next generation are based on these fitness values. The following fitness calculation is an extension of our approach in [9] that was developed for channel routing of integrated circuits.

First, two functions $F_1(p)$ and $F_2(p)$ are calculated for each individual $p \in \mathcal{P}$ according to equations (1) and (2).

$$F_1(p) = \frac{1}{n_{row} + n_{column}} \tag{1}$$

where n_{row} = number of rows of p and
n_{column} = number of columns of p.

$$F_2(p) = \frac{1}{\sum_{i=1}^{n_{ind}} (l_{acc}(i) + a * l_{opp}(i)) + b * v_{ind}} \tag{2}$$

where $l_{acc}(i)$ = net length of net i of net segments according to the preferred
direction of the layer,
$l_{opp}(i)$ = net length of net i of net segments opposite to the preferred
direction of the layer,
a = cost factor for the preferred direction,
n_{ind} = number of nets of individual p,
v_{ind} = number of vias of individual p and
b = cost factor for vias.

In order to assure that the area minimization, i.e. the sum of rows and columns of p, predominates the net length and the number of vias, the fitness $F(p)$ is derived from $F_1(p)$ and $F_2(p)$ as follows:

Assume that $(p_i, ...p_x, ...p_j)$ are individuals with the same area, i.e. the same value $F_1(p)$. These individuals are arranged in an ascending order according to $F_2(p)$. Then p_i is the individual with the lowest value $F_2(p)$ in this group ("worst individual with this area"). Its fitness value $F(p_i)$ is defined by

$$F(p_i) = F_1(p_i). \tag{3}$$

The individual p_j has the highest value $F_2(p)$ in this group ("best individual with this area"). Let $F_1(p_{j+1})$ be the F_1-value of the next ("better") group with one row or column less. The fitness $F(p_j)$ is calculated as follows:

$$F(p_j) = F_1(p_{j+1}) - \frac{\Delta F_1}{j - i + 1} \tag{4}$$

where $\Delta F_1 = F_1(p_{j+1}) - F_1(p_j)$.

Now $F(p_x)$ of the remaining individuals of this group can be calculated relative to their F_2-values between the lower bound $F(p_i)$ and the upper bound $F(p_j)$:

$$F(p_x) = F(p_j) - \frac{\Delta F * (F_2(p_j) - F_2(p_x))}{\Delta F_2} \tag{5}$$

where $\Delta F = F(p_j) - F(p_i)$ and
$\Delta F_2 = F_2(p_j) - F_2(p_i)$.

After the evaluation of $F(p)$ for all individuals of the population $\mathcal{P}$ these values are scaled linearly as described in [5], in order to control the variance of fitness in the population.

3.4 Selection Strategy

The selection strategy is responsible for choosing the mates among the individuals of the population $\mathcal{P}_c$.

According to the terminology of [5], our selection strategy is stochastic sampling with replacement. That means any individual $p_i \in \mathcal{P}_c$ is selected with a probability

$$\frac{F(p_i)}{\sum\limits_{p \in \mathcal{P}_c} F(p)}$$

The two mates needed for one crossover are chosen independently of each other. An individual may be selected any number of times in the same generation.

3.5 Crossover Operator

During the crossover, two individuals are combined to create a descendant. We developed a crossover operator that gives compact, high-quality routing structures of these two individuals an increased probability to be transferred intact to their descendant. Let p_α and p_β be copies of the mates (Figure 4 (a,b)) and p_γ be their descendant.

First, the direction of a cut line (vertical cut column x_c or horizontal cut row y_c) is randomly chosen [1].

From the pin occupation list of the vertical boundaries of the switchbox a pin combination (located on the same horizontal row) is randomly chosen to determine the position of the cut row in each parent p_α and p_β.

The individual p_α transfers its routing structure to p_γ which is

- located on (x_α, y_α, z) with $1 \leq x_\alpha \leq x_{ind\alpha}$ ($x_{ind\alpha}$ = number of columns of p_α), $1 \leq y_\alpha \leq y_{c\alpha}$ ($y_{c\alpha}$ = position of the cut row in p_α), $1 \leq z \leq 2$ and
- not cut by the cut row $y_{c\alpha}$.

Accordingly, p_β transfers to p_γ the uncut connections located on (x_β, y_β, z) with $1 \leq x_\beta \leq x_{ind\beta}$, $y_{c\beta} < y_\beta \leq y_{ind\beta}$ and $1 \leq z \leq 2$ (see Figure 4 (c,d)).

Note that the connections of p_α and p_β cut by $y_{c\alpha}$ and $y_{c\beta}$ are traced until their next Steiner point or pin is reached and not transferred into p_γ.

Assume that the parts of p_α and p_β which have to be transferred into p_γ contain columns not occupied by any vertical segments anymore. Then the number of columns of p_α and p_β are decreased by deleting these empty columns and the sizes of p_α and p_β are shrunk accordingly.

The initial number of columns $x_{ind\gamma}$ of p_γ is obtained by extending p_α and p_β with additional columns to achieve a vertical accordance of pins with the pin occupation list of the switchbox [2].

The routing of the open connections in p_γ is done as follows: Let $\mathcal{N}_\alpha$ be the set of all Steiner points or pins which are end points of a cut segment in p_α. Accordingly, let $\mathcal{N}_\beta$ be the set of these points in p_β. If $\mathcal{N}_\alpha$ contains more than one point of the same net, these points are connected with each other in a random order by our random routing strategy (see Section 3.2). Except for one randomly chosen point, all points of this net in $\mathcal{N}_\alpha$ are now deleted. The same "inner routing" in $\mathcal{N}_\beta$ is performed. As a result, $\mathcal{N}_\alpha$ and $\mathcal{N}_\beta$ do not contain more than one point per net. These points in $\mathcal{N}_\alpha$ are now selected randomly and compared with all points in $\mathcal{N}_\beta$. If a point of the same net is found in $\mathcal{N}_\beta$, both points are connected by means of our random routing (see Figure 4 (e,f)).

If the random routing of two points does not lead to a connection within a certain number of extension lines, the extension lines are deleted and the switchbox is extended at a random position x_{add} with $1 \leq x_{add} \leq x_{ind\gamma}$. If the repeated extension of the switchbox also does not enable a connection, p_γ is deleted entirely and the crossover process starts again with a new random cut row y_c (or cut column x_c) applied to p_α and p_β.

---

[1] In the following explanation of the crossover operator, we assume the direction of the cut line to be horizontal. The crossover procedure resulting from a vertical cut line can be obtained by exchanging the variable x with y and vice versa and the term "row" with "column" and vice versa.

[2] Pins of the horizontal switchbox boundaries which are placed opposite to each other on the same column in the pin occupation list have also to be located in the same manner in p_γ.

195

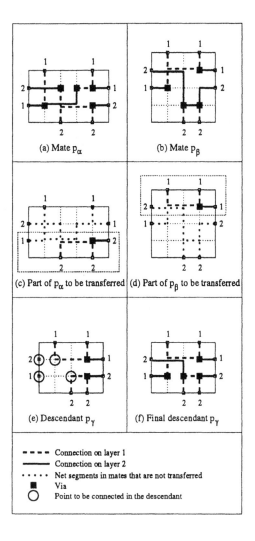

Fig. 4. Crossover of (p_α, p_β) to p_γ.

The crossover process of creating p_γ is finished with deleting all columns and rows in p_γ that are not used for any routing segment.

3.6 Reduction Strategy

Because the population size of a genetic algorithm should be constant, a reduction strategy is necessary to decide which individuals among the current population $\mathcal{P}_c$ and the set of descendants $\mathcal{P}_n$ should survive for the next generation.

Our reduction strategy simply chooses the $|\mathcal{P}_c|$ fittest individuals of $(\mathcal{P}_c \cup \mathcal{P}_n)$ to survive as $\mathcal{P}_c$ into the next generation.

3.7 Mutation Operator

Mutation operators perform random modifications on an individual. The purpose is to overcome local optima and to exploit new regions of the search space.

Our mutation operator works as follows. Define a surrounding rectangle with random sizes (x_r, y_r) around a random centre position (x, y, z). All routing structures inside this rectangle are deleted. The remaining net points on the edges of this rectangle are now connected again in a random order with our random routing strategy.

4   Implementation and Experimental Results

The algorithm, called GASBOR, has been implemented in FORTRAN on a SPARC workstation. The approximate size of the source code is 10,000 lines.

4.1   Measurement Conditions

The performance of the algorithm has been tested on different benchmarks. The routing results of the benchmarks, presented later, are the best results obtained in 10 consecutive executions of the algorithm for each benchmark. All executions are based on an arbitrary initialization of the random number generator. We terminated our program when there was no improvement to the best individual in 100 consecutive generations.

The values of the other parameters are as follows:

$|\mathcal{P}_c|$	= 50
$max\_descendant$	= 30
a	= 1.01 (Equation (2))
b	= 2.00 (Equation (2))
Mutation probability	= 0.002

The same parameter setting is used for all benchmarks.

4.2   Switchbox Routing Results

The routing results of GASBOR for different benchmarks are presented in Table 1. It can be seen that our results are qualitatively similar to or better than the best known results from popular switchbox routers published for these benchmarks.

The detailed results of the 10 executions of GASBOR for each benchmark are presented in Table 2. It must be noted, that the diversity of the routing results of each of the benchmarks is caused by our decision to terminate the program when there is no improvement to the best individual in 100 consecutive generations. As we shall see later, the best results of Table 1 can always be achieved if such a strict abortion criterion is avoided.

In [7, Fig. 6-17], the knowledge-based expert system router "WEAVER" was able to route the so-called switchbox Joo6_17 which is provably unroutable by

Benchmark	System	Rows	Col.	Netlength	Vias
Simple	Lee [7]	7	7	60	11
switchbox	WEAVER [7]	7	7	60	4
	GASBOR	7	7	60	4
Joo6_17	WEAVER [7]	9	11	166	19
	SILK [10]	9	11	166	18
	GASBOR	9	11	164	18
Pedagogical	BEAVER [2]	16	15	396	38
switchbox	PACKER [4]	16	15	406	45
	SAR [1]	16	15	393	31
	GASBOR	16	15	395	31
Dense	WEAVER [7]	17	16	517	31
switchbox	Mighty [11]	18[a]	16	530	32
	SILK [10]	17	16	516	29
	Monreale [3]	18[a]	16	529	32
	SAR [1]	17	16	519	31
	GASBOR	17	16	519	29
Judy	Monreale [3]	17	17	506	32
	GASBOR	17	17	498	32

[a] Additional row at the bottom edge of the switchbox.

Table 1: Benchmark results.

Bench-mark	Generations [a]			Routing result[b]			Run-time[c]
	Min	Max	Avg	Worst	Best	Avg	
Simple switchbox	51	115	69	7/7/ 60/4	7/7/ 60/4	7/7/ 60/4	3
Joo6_17	96	681	398	9/11/ 168/19	9/11/ 164/18	9/11/ 165/18	41
Peda-gogical switchbox	181	1002	517	16/15/ 404/38	16/15/ 395/31	16/15/ 397/32	101
Dense switchbox	198	1192	327	18/16/ 529/32	17/16/ 519/29	17/16/ 520/31	97
Judy	161	998	316	17/17/ 506/32	17/17/ 498/32	17/17/ 500/32	78

[a] Number of generations when the best individual appears in the population.
[b] Routing results given as rows/columns/netlengh/vias.
[c] Average CPU runtime in minutes.

Table 2: Detailed results of the 10 executions of each benchmark.

traditional algorithms. As is evident from Table 1, our algorithm yields a better result than both WEAVER and SILK for this benchmark. Figure 5 shows our routing solution.

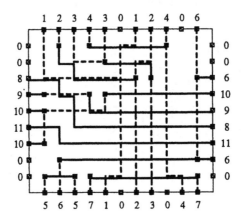

Fig. 5. Our routing solution of Joo6_17.

4.3 Effect of Random Number Generator

Since the methodology of our algorithm is probabilistic, it is important to investigate the effect of the initialization of the random number generator on the routing results.

We conducted an experiment regarding the robustness of our algorithm, i.e., we investigated the impact of the seed of the random number generator on the routing results. The effect of the different initializations of the random number generator on the number of generations needed to achieve our best results was also investigated. We executed our program 1000 times with different initializations of the random number generator to route the switchbox Joo6_17. As soon as the algorithm produced an individual equal to the best one of Table 1, the execution was stopped and the number of generations was noted. The distribution of this number of generations in 1000 program executions is plotted in Figure 6.

Similar results were reached using the other benchmarks of Table 1.

From the experiment we conclude that the initialization of the random number generator affects only the runtime. Our routing results can be achieved with any initial seed of the random number generator.

5 Summary

We presented a GA that is capable of handling the routing problem of switchboxes. Due to its ability to overcome local optima, our GA always achieves the

Frequency

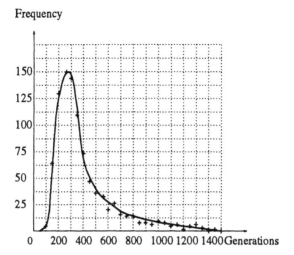

Fig. 6. Distribution of the number of generations needed to achieve the best result of Burstein's difficult channel.

same results for a given routing problem independently of the initialization of the random number generator. Thus, it can be considered as robust.

As can be seen from Table 2, the runtimes of our algorithm on some benchmarks may be unacceptable. However, due to the inherent parallelism of GAs we are optimistic about reducing the runtimes by implementing a parallel version of our algorithm.

Further investigations are needed to measure the performance of the algorithm for larger switchbox problems. Initial examination suggests that an exponential relationship exists between the CPU runtime and the size of the switchbox routing problem.

In developing this algorithm our suggestions for other applications of GAs in the physical design process of integrated circuits are:

- The representation scheme of a layout problem in a GA should be a three-dimensional, problem-specific representation rather than a one-dimensional binary string (as is common in traditional applications of GAs [5]). Our scheme ensures that high quality parts of the layout structure are preserved as high-fitness building blocks and transferred intact with an increased probability in the next generation.
- The genetic operators of a GA in the physical layout design should be adapted to the specific layout problem.
- A sufficient diversity within the population (including the initial population) is crucial to the robustness of a GA for a design problem.
- The ability to overcome local optima is improved by separation of the crossover and the mutation procedures.

From our results we believe that genetic algorithms are promising tools for solving optimization problems in the physical design process of integrated circuits.

References

1. Acan, A., Ünver, Z.: Switchbox Routing by Simulated Annealing: SAR. IEEE International Symposium on Circuits and Systems **4** (1992) 1985-1988
2. Cohoon, J. P., Heck, P. L.: BEAVER: A Computational-Geometry-Based Tool for Switchbox Routing. IEEE Trans. on Computer-Aided Design **7** No. 6 (1988) 684-697
3. Geraci, M., Orlando, P., Sorbello, F., Vasallo, G.: A Genetic Algorithm for the Routing of VLSI Circuits. Euro Asic '91, Parigi 27-31 Maggio (1991) 218-223
4. Gerez, S. H., Herrmann, O. E.: Switchbox Routing by Stepwise Reshaping. IEEE Trans. on Computer-Aided Design **8** No. 12 (1989) 1350-1361
5. Goldberg, D. E.: Genetic Algorithms in Search, Optimization, and Machine Learning. Reading, MA: Addison-Wesley Publishing Company (1989)
6. Goldberg, D. E.: Genetic and Evolutionary Algorithms Come of Age. Communications of the Association for Computing Machinery (CACM) **37** No. 3 (1994) 113-119
7. Joobbani, R.: An Artificial Intelligence Approach to VLSI Routing. Boston, MA: Kluwer Academic Publishers (1986)
8. Lee, C. Y.: An Algorithm for Path Connections and its Applications. IRE-Transactions on Electronic Computers (1961) 346-365
9. Lienig, J., Thulasiraman, K.: A Genetic Algorithm for Channel Routing in VLSI Circuits. Evolutionary Computation **1** No. 4 (1994) 293-311
10. Lin, Y.-L., Hsu, Y.-C., Tsai, F.-S.: SILK: A Simulated Evolution Router. IEEE Trans. on Computer-Aided Design **8** No. 10 (1989) 1108-1114
11. Shin, H., Sangiovanni-Vincentelli, A.: A Detailed Router Based on Incremental Routing Modifications: Mighty. IEEE Trans. on Computer-Aided Design **6** No. 6 (1987) 942-955

The Calculus of Self-Modifiable Algorithm Based Evolutionary Computer Network Routing

Davika Seunarine* and Eugeniusz Eberbach**

Jodrey School of Computer Science, Acadia University
Wolfville, Nova Scotia, Canada B0P 1X0
davika.seunarine@acadiau.ca
eugene.eberbach@acadiau.ca

Abstract. The Calculus of Self-Modifiable Algorithms (CSA) is a universal approach to parallel and intelligent system. Its aim is to integrate different styles of programming and is applied to different areas of future generation computers. Potential applications of CSA include expert systems, machine learning, adaptive systems and many others. The problem of route optimization in computer networks is identified as a task that requires some sort of cost-driven solution that allows for the computation of paths in a network based on experience and inference. The Calculus of Self-Modifiable Algorithm is used to do the specification of this problem by modeling a system of machine learning algorithms that learn proper routing techniques for a particular computer network by incorporating an apportionment of credit system and various rule discovery concepts similar to the learning techniques used in evolutionary computing and symbolic learning.

1 Introduction

The problem of routing data in a communications network involves finding the shortest path between two given points in the network. In such networks, as shown in [17], traditional algorithms like the well known centralized and distributed routing algorithms function well except when faced with multiple and rapid link failures. The basic reason why these conventional algorithms fail is because of the stochastic nature of the traffic pattern and topological changes which lead to the problem of informing all nodes and allowing for immediate update of routes. When such changes take place on the network, at any point in time, a given node knows the accurate state of only its immediate neighbors and the status of all distant paths and links are considered outdated. The calculation of alternative routes at any point in time should therefore be dependent upon a decision system that can decide upon the accurate paths to take based on the

---

* currently a member of scientific staff at Bell-Northern Research, Ottawa, Ontario,
   Canada K1Y 4H7
** research partially supported by a grant from the Natural Sciences and Engineering
   Research Council of Canada, No. OGP0046501

status of neighboring nodes and past history. It therefore requires a decision system that can learn from past experiences and adapt to future changes of network conditions.

Self-Modifiable Algorithms, as specified in [7, 9, 11], are mathematical models of programs with the ability to modify their behavior based on their experiences. The concepts of the CSA have close roots with classic computer science: logic programming, mathematical models of programs, fixed point theory, production rules and others. The CSA is a unifying tool that specifies a general theory which can be used to model various aspects of AI. Production systems have most often been used in AI research to represent the knowledge space of a system and research in production systems have been diverse with a wide range of applications. In [15], rule-based production systems are identified as candidates for the exploitation of general purpose machine learning algorithms and from this idea the concept of *classifier systems* evolved. Such system incorporated learning mechanisms similar to the "survival of the fittest" concept used in Darwin's theory of evolution.

The paper addresses the above mentioned routing problem with respect to a packet switch data communication network from an evolutionary cost driven programming point of view , incorporating the concepts of the classifier systems and other machine learning techniques from symbolic learning in order to learn proper routing techniques. The modeling and implementation is done using the Calculus of Self-Modifiable Algorithms to specify the cost-driven program that responds to changes in the environment as changes occur. The implementation of the such cost-driven program has been done in SEMAL, which is a representative of a new cost language programming paradigm specified in [11] and derived from the Calculus of Self-Modifiable Algorithms.

The layout of the paper is as follows: Section 2 discusses conventional routing algorithms, identifying some of the strengths and weaknesses of these algorithms. Section 3 presents inference-based learning and evolutionary approach in AI research. Section 3 examines the Calculus of Self-Modifiable Algorithms and shows how it can be used to specify evolutionary learning concepts, and Section 5 presents the specification and working of a CSA-based computer network routing algorithm that incorporates some of the conventional learning techniques and an evolutionary approach.

2 Routing Algorithms

In [1], routing algorithms are described as decision making algorithms that are responsible for deciding which outgoing links a particular node of a network should route incoming as well as generated packets that are destined for a particular destination node. This function takes place in the Network and Transport Layers in the Reference Model of the Open Systems Interconnection (OSI) proposed by the International Standards Organization(ISO).

This decision making process should perform the routing function at minimum cost of delay to data packets. Adaptive routing algorithms aim to make such decisions based on the current state of the network in terms of traffic patterns and network topology.

A good routing algorithm should increase the throughput of the network when traffic load on the network increases, with packets experiencing no change in delays and to decrease the delay per packet when the load on the network decreases so as to make full use of the network capacity. The routing algorithm should therefore operate effectively for both high network loads as well as low network loads so as to keep the average delay per packet as low as possible.

The algorithm therefore needs to make new decisions when current paths become congested as traffic is to be redirected elsewhere in the network. Some algorithms attempt to address these factors while others do not; the need to address these factors basically depends on the stability of the network. For networks that are highly unstable it is important that the routing decisions be updated on a regular basis, while for highly stable networks the routing decisions are updated on a less frequent basis. The various types of network routing algorithms can therefore be classified as:

Non Adaptive Routing - These types of algorithms do not base their decisions on measurements or estimates of the current state of the network. The choice of route to take in order to reach at a particular destination is computed once and never changes.

Adaptive Routing - These types of algorithms attempt to change their routing decisions so as to reflect topological changes and traffic conditions on the network. They attempt to make new routing decisions that would guide network traffic around the areas of congestion in the network.

Routing algorithms can also be classified into the categories of *Isolated, Distributed, Centralized,* and *Hybrid* routing. These techniques basically differ in the way routing paths are computed and implemented. The following section gives a brief overview of these categories.

2.1 Classification of Routing Algorithms

Isolated Routing - These algorithms run individually on each node and each node makes its routing decisions based on purely locally available information.

Distributed Routing - in the distributed method of path computation nodes exchange routing information and the routing decision at each node is based on a combination of locally available as well as exchanged information. These algorithms basically work on the fact that if neighbor Y of node X has estimated distance to destination D of X_D, and the length of link $X - Y$ is l_{xy} then node $X's$ estimate to destination D via neighbor Y is $l_{xy} + X_D$. Most distributed routing algorithms are adaptive, however any of these algorithms can be termed non adaptive if they are used only once at network setup time to determine shortest paths and never again.

Centralized Routing - with this type of routing, a node, upon receiving topological broadcast information from each other node in the network, construct the routing table based on an internal graphical representation of the network. This may take place at each node, thus each node created its own routing table, or it may take place at a central routing control center and the routing table information is then transmitted down to each node.

Hybrid Routing - this technique is a combination of centralized and isolated routing. Most hybrid routing algorithms are adaptive.

2.2 Problems with Conventional Routing Algorithms

The various categories of routing algorithms which were presented in the previous sections all have strengths and weakness under different network conditions.

With the centralized method of path computation, although the routing computation works to produce optimal paths based on available information, this information is outdated because of the time lag involved in getting status information from each node. Although there is no looping problems, the overhead of message requirements and memory usage is high as each node must gather information about the entire network topology.

The initial implementation of the distributed routing algorithms has proved to be very inefficient in the face of link cost increase and node failures. Further improvements in these algorithms [4] worked to solve this problem but with the requirements of additional information at each node.

Isolated routing algorithms have been shown to be short- sighted in making their routing decisions because no consideration is given to the global network structure. Although these algorithm work on up-to-date information on neighboring nodes, this information is not sufficient to create optimal paths.

The basic reason why these conventional algorithms fail to function efficiently under certain conditions is because of the stochastic nature of the traffic pattern and topological changes which leads to the problem of informing all nodes and requires immediate update of routes. Most routing algorithms, when faced with link or node failures has to delay traffic on the network until the failure is detected and route updating is completed by all nodes and only then can packets be routed to a new path [17, 18]. Some algorithms lead to the creation of loops in the paths and the *counting to infinity* problem when all nodes are not immediately updated with topological changes [3, 6, 17].

Another major reason for the above mentioned drawbacks is that each node has to operate with incomplete information on the state of the rest of the network. Most often the routing algorithm would have accurate knowledge of only neighboring links. Any other information as to the state of the network is out of date and therefore unreliable for use in calculating new routes. This problem therefore requires a decision system that can learn from past experiences, deals with incomplete information, is dynamic, and can make fairly good decisions when faced with such problems. Such systems should be able to effectively combine the merits of the various routing techniques as described in the previous sections, with some method of determining which technique is more suited to a particular network environment. The hybrid method of routing information attempted to combine the merits of isolated and centralized routing, but this decision making process is still static in the sense and may work well for only some network conditions. There is therefore a need to establish some sort of system where the routing techniques can evolve with changing network conditions so that alternative paths can be chosen by an efficient means of calculation for the given network conditions. Such system would therefore require some sort of intelligence and thus incorporate some sort of machine learning techniques.

3 Inference-Based Learning and Evolutionary Algorithms

In recent years rule based production systems have been applied to a wide variety of application domains so as to exhibit some sort of intelligence. However, most of them do not have learning capabilities because they do not acquire nor update their knowledge dynamically. A machine is said to have learning capabilities when it improves its performance at a given task without any human intervention. In order to exhibit more intelligence, a system must be able to reason and learn from the experiences that it encounters and modify the rules that it works with, so as to form better rules and principles. The rule space is usually a set of possible rules or heuristics that the system makes use of in inferring suitable solutions to a problem. Traditional AI attempt to do this by uses the concept of **Induction** which involved *Dropping Condition, Adding Options, Turning Constants into Variables* , and **Deduction** which is basically the reverse of the rules of induction the aim being to make rules more constricted and therefore more specialized in solving a particular problem.

When new rules are created there must be some way of evaluating the merits of the new rules to find out how useful they are to the system. Traditional AI did not attempt to address this factor, and in 1978 Holland proposed a message passing rule based system that exhibited the capabilities of an intelligent systems which incorporated such an evaluation system. This system used a method of rule discovery and inference that was based on the concept of evolution, as it takes place in nature, where new rules are evaluated and must compete with others in order to prove their usefulness. This system was called a *classifier system* [14, 15].

Classifier systems are a special type of rule-based production systems that was first introduced by Holland and Reitman in 1978. In such systems many rules can be active at the same time so information can be processed in a parallel way. Classifier systems consist of a set of rules(classifiers) that manipulate an internal *message list*. The precondition of each classifier specifies a pattern that matches messages in a message list and the postcondition specifies messages that are to be posted to the message list if the classifier is to be fired. Interaction with the environment takes place by the use of *detectors* which post detector messages to the message list along with a set of *effectors* which generate external actions in response to posted messages. The rules in the classifier systems are designed to be modified by genetic-based learning algorithms.

Learning in Classifier Systems

Three factors work together to provide learning in classifier systems. They are:

1. *Competition to post messages :* Classifiers compete to post messages, thus encouraging the activation of only some of the classifiers satisfied at any particular point. In [15], a bidding process in suggested where classifiers which are matched at any particular instance must bid to become active and only the highest bidders are allowed to become active and post messages to the message list.

2. *The Bucket Brigade Algorithm :* This mechanism seeks to allocate strength (or credit) to a classifier which tend to produce useful system behavior. There are many ways of doing the actual implementation of this system. The approach most often taken is based on a reward or pay-off scheme where each time the classifier system produces actions that the environment considers as desirable, the classifier system receives a positive reward from the environment.

3. *Rule Discovery Algorithm :* New messages are created based on the existing classifiers in the system. In [14], Holland described the use of genetic algorithms to do such rule discovery. The basic algorithm is as follows:
   - Select some classifiers to act as parents based on their strength or fitness.
   - Make copies of these classifiers and perform genetic crossover and mutation to produce new classifiers.

   These resulting classifiers are placed in the classifier list and they enter into competition with the other classifiers.

Various analysis and simulation have shown that such rule-discovery algorithms, when used together with the bucket brigade algorithm and the competition mechanisms provide useful ways to search and discover useful classifiers.

Genetic algorithms, which are used for the rule discovery phase of the Classifier-System, are a family of adaptive search procedures that have been extensively analyzed in [14].

The Calculus of Self-Modifiable Algorithm is described in the following section as a model that can do the specification of an adaptive system similar to in concept to the classifier system, but with the capability to incorporate various other machine learning techniques.

4 The Calculus of Self-Modifiable Algorithms

The Calculus of Self-Modifiable Algorithms (CSA), as described in [8-13], was designed to be a universal theory for intelligent and parallel systems, integrating different styles of programming and applied in different domains of future generation computers. The use of artificial intelligence in future generation computers requires different forms of parallelism, learning, reasoning, experience and knowledge acquisition for which there have been a considerable lack of theoretic models proposed. The CSA proposes such a theory, by introducing a mathematical model of programs with the ability to modify their behaviors based on past experience and thus accomplish most of the outlined requirements for future generation computers. CSA can therefore be used to model adaptive computer systems (programs, processes and architectures) and such "programs" are called Self-Modifiable Algorithms (SMA). The SMAs are mathematical models of programs with the ability to remember history of their past realization, and therefore they possess a certain degree of self-knowledge and this, together with its built-in optimization mechanism provide goal directed modifications of programs or algorithms so that completely new forms of algorithms can be built to achieve the goal with minimal cost. SMAs are based on mathematical models of programs; with each component having a frame-like structure and with the use of the fixed-point theory to describe the input-output behavior, and thus, the history of computation. The entire SMA structure is dynamic and changeable because of the modification transits that are incorporated in it. The modifications either change the existing transits' preconditions, actions, postconditions, or costs or create new transits. Thus, unlike a conventional program where given a particular set of input data the program react in a deterministic way and always provide the same output result, a SMA has the ability to produce results based on learning and experience and therefore, given a particular set of input data, the system responds differently, with the aim of providing better solutions. The CSA theory can be applied to a wide range of research areas, including expert systems, machine learning, fault tolerant computing and neurocomputing [8, 19, 20, 21, 22]. CSA has been proposed to be a suitable tool to investigate fifth generation "knowledge-based" computers and sixth generation "intelligent" computers. It seems that it is an appropriate tool for machine learning, i.e., for learning by analogy, from examples, from observation and by discovery. The cost language SEMAL, for which further information can be found in [12], is based on the CSA. The following sections describe an outline of the theory behind the SMA.

4.1 Formal Environment for Self-Modifiable Algorithms

Extended Boolean Algebra Formulas

In [8], the definition of Extended Boolean Algebra Formulas (EBA-formulas) is given as a class of logic formulas which can be expressed by the following Backus-Naur Form syntax:

$$\Phi ::= 1 \mid 0 \mid - \mid X \mid P \mid (\Phi) \mid \forall x.\Phi \mid \exists x.\phi \mid \neg\Phi \mid \Phi \vee \Phi \mid \Phi \wedge \Phi \mid \Phi; \Phi \mid \Phi \leadsto \Phi \mid \Phi^n$$

EBA-Formulas are used to specify predicates that represent control states, preconditions, activities and postconditions of transits which are the basic programming units of the SMA. EBA-Formulas are similar to predicate calculus, the main differences are:

- Three-valued logic(True, False and Unknown represented by 1,0 and - respectively).

- X are logic variables.
- P is the set of n-ary relations, where $\forall p \in P$ p is the name of the n-ary relation $p(t_1, t_2, ..., t_n)(n \geq 0)$, and $(t_1, t_2...t_n)$ are terms. Terms include constants, variables and functions.
- $\forall x.\Phi, \exists x.\Phi$ are quantified EBA-Formulas with $\forall$ being the **universal quantifier** and $\exists$ being the **existential quantifier**.
- $\wedge, \vee, \neg$ are **conjunction, disjunction** and **negation** respectively.
- In addition to Conjunction, Disjunction and Negation operators there are new operators as follows :
    1. **String** dynamic temporal-like operator(;) - Describes the temporal-like precedence of EBA-formulas and therefore can be used to represent the execution sequence of a system.
    2. **Verification of Matching** modus ponens based operator($\rightsquigarrow$) - Matches control states and transits to find potential candidate transits for execution.
- Φ^n is EBA-formula with **capacity** n. Capacities denote how many transits can use (share) a given formula in parallel.

Updating Condition Operators

The **Forward**, $\triangleright$, and **Backward**, $\triangleleft$, **Updating Condition** operators are used to create new control states of SMA given the state before and the matching transit (forward updating), or given the state after and the matching transit (backward updating).

SMA-Net

This provides the means of procedural and data abstraction in the CSA. It therefore provides the means of building the algorithm and high-level transits from lower-level elements (transits) using basic programming operators.

- *The lower-level elements* - These are atomic transits that are made up of *Preconditions, Activities, Postconditions* and *Costs*. Execution of a transit takes place in three steps: first the Precondition is set to true and the Activity and Postcondition are false, then the Activity becomes true and the Precondition and Postcondition are false and then finally the Postcondition is to true and the Precondition and Activity are false.
- *The Operators* - These are the Nondeterministic Choice, $\uplus$, General Choice, $\sqcup$, Sequential Composition,$\circ$, Parallel Composition,$\|$, and General Recursion, $\Re$, operators. These operators are used to connect atomic elements (or transits).

By a SMA-net we therefore mean the following system:

$$(X, \{pre, act, post\}, \{\uplus, \sqcup, \circ, \|, \Re\}, \{\bot, \top, \varepsilon, \gamma\})$$

X - is a set of atomic elements of the net which are called transits. $X = \Phi \times \Phi \times \Phi$, where Φ are EBA-Formulas. Therefore, each X is of the form (ϕ_1, ϕ_2, ϕ_3), where ϕ_1 is the Precondition, ϕ_2 is the Activity and ϕ_3 is the Postcondition of the transit. SMA-Expressions, $Exp(X)$, are defined recursively as:

$$\forall x \in X \cup \{\top, \bot, \varepsilon, \gamma\} \; x \in Exp(X)$$

$$\forall x, y \in Exp(X) \; x \uplus y \in Exp(X), x \sqcup y \in Exp(X),$$

$$x \circ y \in Exp(X), x \parallel y \in Exp(X), x \, \Re \, y \in Exp(X)$$

$\uplus, \sqcup, \circ, \parallel, \Re$ are binary operators with their meaning informally described as follows:

$\uplus$ - is the Nondeterministic choice operator and is used when the preconditions of more than one transits are true at the same time, and therefore is used in the situation where any of the actions can be fired. In the case of routing in computer networks, it can be used in the case where there are two different routing techniques to be taken at a particular point in the network.

$\sqcup$ - is the General Choice operator and the method of usage, $(a \sqcup b)$, means that if a is true then a is executed, or if b is true then b is executed (usually the preconditions of a and b are not true simultaneously).

$\circ$ - is the Sequential Composition operator and is used to describe successive events.

$\parallel$ - is the Parallel Composition operator or Intersection operator. It's definition specifies the control checks that must occur before two transits can be executed in parallel.

$\Re$ - is the General Recursion operator which defines one transit as a function of other transits (including possibly itself) and SMA-net operators.

$\bot, \top, \varepsilon, \gamma$ are special elements of the SMA-Net and are defined as:

$\bot$ - is the *zero* element of the net and can be considered as a deadlock conditions or a transit that never fires because its precondition is false.

$\top$ - is the largest element of the net and it incorporates all possible transits. Its precondition is therefore always true.

ε - is sequential unity which is invisible in a sequential execution, ie., $a \circ \varepsilon = a$.

γ - is parallel unity which is invisible in a parallel execution, ie., $a \parallel \gamma = a$.

Cost System

The Cost system is used for SMA learning and adaptation. Each transit and SMA-operator are assigned a cost and the cost of the SMA equations that are built is a composition of this cost and the cost of the SMA-Net operators. SMA learning is achieved by finding the minimum of costs for a particular subset of transits that provide the cheapest solution. This process is done by the SMA inference engine optimization process and therefore the process of finding the best solution involves the mathematical process of finding a minimum. The cost value has different interpretations and can be problem specific. For example, it can represent weights in the neurocomputing sense, time of execution penalty, or in the evolutionary programming case it can represent the fitness of the transit in contributing to a useful solution.

4.2 Self-Modifiable Algorithm

Specification of a Self-Modifiable Algorithm SMA over a SMA-net includes:

- **Control States S** - Initial Control State (σ) and Final Control State (ω) are specified at the beginning; other control states are created during running of the algorithm.
- **Transits T** - Consist of Data D, Actions A, Modifications M. Data are passive, Actions act on Data, and Modifications act on both Data, Actions, and Modifications. Each transit consists of Precondition (pre), Activity (act), Postcondition (post), and Cost (cost). Preconditions, activities, and postconditions have the form of EBA-formulas, and costs are represented by real (or so-called beyond-real) numbers.

4.3 Inference Engine of Self-Modifiable Algorithms

Unsimilar to GPS or Strips-like problem solvers, the SMA Inference Engine finds not a path, but an algorithm, thus the SMA-equations contain alternative solutions, which are connected by general choice operator, $\sqcup$, nondeterministic choice operator, $\uplus$, and parallel solutions which are connected by parallel composition operator, $\|$. One of the basic properties of each transit is its cost. The optimization process finds the best algorithm or solution to the problem by finding the cheapest SMA that leads from the initial to the terminal control states. The process of problem solving depends on the complexity of the problem space itself; for easy problems the entire algorithm is built up to the end of the program, for complex problems heuristic searches are used to find the best solution a few steps ahead based on the SMA cost function. The Inference Engine works in three phases : the *select, examine* and *execute* phases.

The *select* phase involves the verification of matching of the SMA-transits and the building of the equations. The inference engine builds sets of equations k steps forward or backward, where $k = 1, 2, 3..., \infty$, where $k = \infty$ means to "the end" or to "the beginning" of the algorithm, (depending on whether forward or backwards matching was used respectively) thus building the tree of control states starting from the current control state x and moving to a new state $x + 1$ by making use of the transits that match the current control state. Such process continues for k steps until level $x + k$ is reached.

In the *examine* phase the inference engine selects the optimal program from the expanded control states based on the cost of the least fixed-points solutions of equations, and this program is executed in the execute phase. The examine phase of the inference engine therefore involves finding fixed point solutions of the SMA-equations that were built and the optimization of such equations by removal of non-determinism from the equations. The fixed point solution contains redundant threads of transits and the optimization process involves removing the non-determinism from the equation by using the cost of transits and operators to select the best algorithm to solve the problem. Hence for the least-fixed-point, t, of the set of equations, if $\varphi(t_1)$ represents the cost of transit

t_1, then this process involves the application of the *erasing homomorphism* $h(t)$ as follows:

$$h(t) = \begin{cases} t & \text{if it does not contain } \uplus \\ t_1 & \text{if } t = t_1 \uplus t_2, \text{ and } \varphi(t_1) = min(\varphi(t_1), \varphi(t_2)) \\ t_2 & \text{if } t = t_1 \uplus t_2, \text{ and } \varphi(t_2) = min(\varphi(t_1), \varphi(t_2)) \end{cases}$$

The execution of the optimized program takes place in the *execute* phase. The inference engine then starts the process from the select phase again and this goes on until either no more control states can be generated or the terminal control (or initial) state is reached.

4.4   Evolutionary Programs as Self Modifiable Algorithms

Evolutionary programs can be viewed as a hierarchy of programs, each with varying knowledge about a specific problem. These programs vary in their usefulness and accuracy towards solving a specific problem. The fitness measure used in evolutionary programs is a measure of how close the actual solution, which is provided by the program, is to the required solution.

SMAs can be used to specify the cost driven evolution programs from a weak solution to a strong solution. In the general case, the cost of a transit in the SMA can be a combination of the complexity of the transit and the accuracy of the transit towards solving the problem. The problem at hand must be well defined in terms of control states and transits (with EBA-formulas in Preconditions, Activities and Postconditions) and the state of the system at a future time is controlled by the choice of SMA-modifications that modify transits with the aim of creating transits with lower costs that are better fit to solving the problem at hand.

A weak solution to the problem would therefore be a population of actions that perform the task but not necessarily in the most efficient way. A further level of modification would take place to actions where the Preconditions, Activities, Postconditions and Costs are modified to include more accurate knowledge of the problem. By so doing the search for an efficient program would move to the optimal region of the search space. Thus, by the use of the CSA we expect that it would be possible to build programs that learn to solve a particular problem. Such approach may be possible to model the learning process that is necessary to learn proper routing techniques in a communications network as outlined in Section 2.

The following section presents the specification of the routing problem in the CSA approach.

5   The Self-Modifiable Algorithm Specification of an Evolutionary Routing

The approach taken is to model a system similar to the evolutionary algorithm based Classifier System which has the capability of learning better heuristics based on some initial set of heuristics. The CSA approach to modeling such a system is more general than the classifier system, mainly because the CSA approach allows for the incorporation of various other rule discovery techniques, as well as the fact that the EBA-formula representations of a transit's precondition and postcondition allows for the representation of various heuristics that can normally be represented in AI applications but not necessarily encoded in binary format.

5.1 Problem Definition

Consider the simple network layout shown in Fig.1. Each node must be able to route information to every other node on the network along the most efficient path. If we simplify the problem by looking at the routing problem from all nodes to one particular destination DES, we can therefore work on the assumption that the final state of the network is a $SINK$ tree which is a directed graph, G, from all nodes, N, to the given destination DES. For simplicity, assume that the cost of link $X \rightarrow Y$ is the same as cost of link $Y \rightarrow X$ and therefore if link $X \rightarrow Y$ increases in cost, so too does link $Y \rightarrow X$.

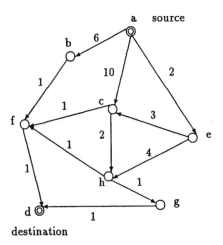

Fig. 1. Network Layout

Let the length of the path from any node X to destination node DES on G be $P(X, DES)$, then the net cost of routing information as specified by G from N nodes to destination DES is therefore :

$$\sum_{i=1}^{N} P(X_i, DES)$$

In order for information to be routed efficiently along the network, this summation must be the minimum of all possibilities on the network. In the CSA approach, it is therefore the responsibility of the CSA inference engine to maintain this minimum and therefore work to maintain the following inequality :

$$\sum_{i=1}^{N} P(X_i, DES) <= \epsilon$$

where ϵ is a predefined value specific to the network topology and cost of links. The problem can be basically divided up into two parts:

1. Upon the initial setup of a node on the network, the algorithm should work so as to find the best path for that node to route information to destination DES. This would be done by using the CSA inference engine optimization process on a set of initially selected heuristics. It is assumed that the global network topology is unknown by each node and that each node knows only with accuracy the topological structure of only its surrounding neighbors. The problem therefore comes down to the one of having each node make a decision as to which of its neighbors it should route information to if that information is destined to a particular node. This decision is made solely on locally available information, together with a set of initial heuristics that make decisions based on this information. The heuristics themselves are candidates for change in the CSA-based system, as new heuristics are created and exploited in order to arrive at a set of heuristics that would make good decisions for the specific network structure.

2. Once all nodes are set up, there are still lots of changes that can take place to the network that would make the above inequality false and this is when the optimization process comes into play again. The process of optimization works on the locally available information together with the set of routing heuristics so as to come up with the best choice of alternative paths. The following initiate the second optimization process:

   - $Link X - Y$ increases in cost and link $X - Y$ is on the graph G
   - $Link X - Y$ decreases in cost and $X - Y$ is not on the graph G
   - $Link X - Y$ becomes detached and link $X - Y$ is on the graph G

   These problems can be solved by the same inference engine optimization techniques as described above for the initial setup of a node on the network. In this case the SINK graph G would be better defined, most other nodes would have estimates to the destination DES and the rule base (or transits in the case of CSA) would have a set of heuristics that are better suited to handling link modifications for the specific network environment.

5.2 Overview of the CSA Specification

The CSA would be used solve the routing problem in data networks by making use of a set of well known routing heuristics and using these as a base to learning new and better heuristics. The learning process takes place by making use of genetic algorithms and other inference based learning procedures which encourages the evolution of good routing heuristics which can make feasible decisions based on available information.

Information Available at Each Node

Each node knows with complete accuracy, the status of only its surrounding neighbors. All other information may be outdated. A give node, X, therefore can know the following about its neighbors :

1. How many neighbors it has, $NEIGHBORS(X)$
2. The cost of transmitting data on the direct link to each neighbor Y, $LINK(X,Y).cost$
3. Its current estimated cost to destination node D, $PATH(X,D).cost$

Choice of Initial Heuristics - SMA actions A set of initial heuristics that work well for finding the next node to a particular destination is identified. The following heuristics work well for a particular node, X, for making the decision of routing to node Y under certain network setup:

Random Walk Method - This is the most general heuristic and it guarantees a solution for the routing of data to a destination, but not necessarily via the most efficient means of computation. Hence node X would route to the destination via neighbor Y, chosen randomly, if $LINK(X,Y)$ exist.

Greedy Method - This is a technique that makes a "short sighted" decision as to which next node to route information to. It does not guarantee an optimal solution unless it works in a centralized way as in the Dijkstra's Algorithm. However, it does attempt to make the best decision based on the locally available information and is therefore included in the initial set of heuristics. Hence node X routes to the destination via Y if condition $(LINK(X,Y) \wedge LINK(X,Y).cost < \lambda)$ is fulfilled, where λ is a predefined minimal value.

Distributed Method - The initial set of heuristics also includes a technique from the distributed method of calculations where a node makes a decision to route information via another node based on the second node's estimate to the destination. Note that this estimate may not necessarily be up-to-date and therefore this has to be taken into consideration when making this decision. Hence node X route to the destination via Y if condition $(LINK(X,Y) \wedge PATH(Y,D).cost < \eta)$ is true, where η is a predefined minimal value.

Isolated Method - In this case a node makes a decision to route to a particular node based on the information it knows about that neighbor, for instance the number or outgoing links that neighbor has, and therefore the amount of alternative options to explore. Hence node X route to the destination via Y if condition $(LINK(X,Y) \wedge NEIGHBOR(Y) > \beta)$ is true, where β is a predefined value.

The Inference Engine Processing The inference engine processing is initiated in the first instance upon the need to create the initial routing table and on subsequent instances when the cost of routing information increases. The CSA inference engine works by examining the initial-control-state and doing a forward match of all applicable transits, including actions and learning modifications, for k steps. After expansions of all states and the creation of the SMA equation, the CSA inference engine optimization mechanism is then initiated and works to minimize the cost of the route to a particular destination by selecting the path of lowest cost (thus highest fitness) that was created by the action (or combination of actions) of highest fitness. In the implementation, the fitness values are represented by negative costs and thus increasing the fitness of a particular transit means decreasing the cost of that transit.

The Rule Discovery Mechanisms The CSA modeling of this problem incorporates genetic algorithm type modifications and some other inductive learning techniques in order to introduce new heuristics into the system. These learning techniques are implemented as SMA-modification transit and the initiation of these transits cause the learning process to take place. The rate at which the learning process take place is controlled by the user and is as follows:

- For simple network structures the learning process may be not necessary because a solution can always be found in a reasonable amount of time. This is so because the inference engine works in expanding the equation until the goal state is reached.
- For more complex network structures, the inference engine optimization process takes place for k steps, where the value of k is determined by the user. At the end of each step, the new heuristics are introduced into the system based on the experience that the system encountered in the previous optimization step. These new heuristics are placed in the system in an attempt for better decisions to be made at each point and hence reduce the size of the SMA equation by only considering feasible options rather than all possible options.

5.3 The CSA Specification

We would look at the problem of establishing the initial path from one node, say node A, to destination node D. It is assumes that none of the nodes on the network has initial estimates of distances to the destination. All the necessary information required by the initial set of heuristics are incorporated in the initial-control-state of the algorithm. In the CSA expressions the following conditions are used :

$LINK(X, Y)$ - Directed link $X - Y$ exists
$LINK\text{-}LESS\text{-}THAN(n, X, Y)$ - The length of link $X - Y$ is less than the value
   n

NEIGHBOR-GREATER-THAN(n,X) - Node X has more than n neighbors (outgoing links)

PATH-LESS-THAN(n,X) - Node $X's$ approximate distance to destination D is n

Note that the value of γ that appears in each of the SMA-action transits that follows represent the fitness of that particular action in making useful routing decisions. When a SMA-Action is matched and added to the SMA-Equation in the equations building stage, a portion of this fitness value is subtracted from the action. After the optimization phase, the bucket-brigade-modification, *M1*, then allocates the necessary payoffs to the actions that provided the optimal solution, hence the strengths of these actions are increased. The fitness value really reflects the SMA-cost of the action and since the SMA-Inference Engine works by minimizing cost, then large values of γ represents weak actions and small values represent stronger actions.

Let's assume that initial fitness of all actions is equal, i.e. $\gamma_1 = \gamma_2 = \gamma_3 = \gamma_4 = 0$ (this means cost is minimal). Let STEPS=0, let i++ denote increment of i by 1.

The SMA specification is as follows:

$$SMA = (S, T)$$

$S = \{\sigma, \omega\}$, where

σ = STATE(a) $\wedge$ (STEPS $\neq$ 2) $\wedge$ NET-INF /*INITIAL-CONTROL-STATE */
NET-INF = LINK(a,b) $\wedge$ LINK(a,c) $\wedge$ LINK(a,e) $\wedge$ LINK(b,f) $\wedge$ LINK(f,d)
$\wedge$ LINK(h,f) $\wedge$ LINK(c,h) $\wedge$ LINK(e,c) $\wedge$ LINK(h,g) $\wedge$ LINK(g,d) $\wedge$
LINK(c,f) $\wedge$ LINK-LESS-THAN(3,b,f) $\wedge$ LINK-LESS-THAN(3,a,e) $\wedge$LINK-
LESS-THAN(3,c,f) $\wedge$ LINK-LESS-THAN(3,f,d) $\wedge$LINK-LESS-THAN(3,h,f)
$\wedge$ LINK-LESS-THAN(3,h,g) $\wedge$ LINK-LESS-THAN(3,g,d) $\wedge$ NEIGHBOR-
GREATER-THAN(1,a) $\wedge$ NEIGHBOR-GREATER-THAN(1,c) $\wedge$
NEIGHBOR-GREATER-THAN(1,e) $\wedge$ NEIGHBOR-GREATER-THAN(1,h)

ω = STATE(d) /* TERMINAL-CONTROL-STATE */

$T = \{D, A, M\}$, /* TRANSITS */, where
$D = \{$a,b,c,d,e,f,g,h$\}$ /* DATA */
$A = \{$A1, A2, A3, A4$\}$ /* ACTIONS */
$M = \{$M1, M2, M3, M4, M5$\}$ /* MODIFICATIONS */

DATUM X /* $X \in \{a, b, c, d, e, f, g, h\}$ */
   pre: X
   post: X
   cost: 0

ACTION A1(X,Z) /* RANDOM WALK HEURISTIC */
   pre : STATE(X) $\wedge$ LINK(X,Z) $\wedge$ (STEPS $\neq$ 2)
   act : INCLUDE-LINK(X,Z) $\wedge$ (STEPS++)

post: STATE(Z) ∧ LINK(X,Z) ∧ ((STEPS ≠ 2) ∨ (STEPS=2))
cost: link(X,Z).cost + γ_1

ACTION A2(X,Z) /* GREEDY METHOD */
  pre : STATE(X) ∧ LINK(X,Z) ∧ (STEPS ≠ 2)
        ∧ LINK-LESS-THAN(3,X,Z)
  act : INCLUDE-LINK(X,Z) ∧ (STEPS++)
  post: STATE(Z) ∧ LINK(X,Z) ∧ ((STEPS ≠ 2) ∨ (STEPS=2))
       ∧ LINK-LESS-THAN(3,X,Z)
  cost: link(X,Z).cost + γ_2

ACTION A3(X,Z) /* ISOLATED METHOD */
  pre : STATE(X) ∧ LINK(X,Z) ∧ (STEPS ≠ 2)
        ∧ NEIGHBOR-GREATER-THAN(1,Z)
  act : INCLUDE-LINK(X,Z) ∧ (STEPS++)
  post: STATE(Z) ∧ LINK(X,Z) ∧ ((STEPS ≠ 2) ∨ (STEPS=2))
       ∧ NEIGHBOR-GREATER-THAN(1,Z)
  cost: link(X,Z).cost + γ_3

ACTION A4(X,Z) /* DISTRIBUTED METHOD */
  pre : STATE(X) ∧ LINK(X,Z) ∧ (STEPS ≠ 2)
        ∧ PATH-LESS-THAN(3,Z)
  act : INCLUDE-LINK(X,Z) ∧ (STEPS++)
  post: STATE(Z) ∧ LINK(X,Z) ∧ ((STEPS ≠ 2) ∨ (STEPS=2))
       ∧ PATH-LESS-THAN(3,Z)
  cost: link(X,Z).cost + γ_4

MODIFICATION M1(A) /* BUCKET BRIGADE */
Decrease cost by, say 1 (increase fitness) for actions chosen for optimal solution, and increase costs of others (penalize them) by 1. M1 will match after 2 STEPS of action execution (our arbitrary choice when modifications will be called).
  pre : (STEPS = 2)
  act : BUCKET-BRIGADE-MODIFICATION(A)
  post: GENERATE-NEW-ACTIONS
  cost: 1

MODIFICATION M2(A_x, A_y) /* SPECIALIZATION */
Create a new action A_z whose precondition is the conjunction of preconditions of A_x and A_y. If action is new add it to the set of actions, with the cost being minimum cost of their parents.
  pre : GENERATE-NEW-ACTIONS
  act : SPECIALIZE(A_x, A_y, A_z)
  post: REMOVE-WEAK-TRANSITS
  cost: 1

MODIFICATION M3(A_x, A_y) /* CROSSOVER */
Create new actions A_{z1} and A_{z2} whose preconditions are the crossover of preconditions of A_x and A_y. If action is new add it to the set of actions, with the cost being minimum cost of their parents.
    pre : GENERATE-NEW-ACTIONS
    act : CROSSOVER(A_x, A_y, A_{z1}, A_{z2})
    post: REMOVE-WEAK-TRANSITS
    cost: 1

MODIFICATION M4(A_x) /* MUTATION */
Create a new action A_z from A_x by randomly negating parts of the precondition with the cost being equal to the cost of its parent.
    pre : GENERATE-NEW-ACTIONS
    act : MUTATE(A_x, A_z)
    post: REMOVE-WEAK-TRANSITS
    cost: 1

MODIFICATION M5(A) /* REMOVE UNFIT TRANSITS */
Remove transits with γ exceeding, say 12 - not useful for the system - too small fitness.
    pre : REMOVE-WEAK-TRANSITS
    act : REMOVE-TRANSITS-WITH-COST-LARGER(12,A) $\wedge$
        (STEPS:=STEPS-2)
    post: (STEPS $\neq$ 2)
    cost: 1

5.4 Work of the Inference Engine for k=5 Steps

The inference engine works by selecting the set of transits that match the given network condition, expanding all possibilities of routes and choosing the set of rules that build the cheapest equation that routes information to destination DES. In our case $\Omega = A$, i.e. the Inference Engine ignores costs of data and modifications (treats them as equal 0). The work of the Inference Engine for $k = 5$ steps (2 steps used by actions for routing, and 3 steps by modifications) is shown below. The $k\Omega$-**Procedure** generating solutions will consist of two cycles i=1,2.

i=1 SELECT phase:
Let σ_{+b+x}^{-a-y} denote that STATE(a) has been replaced by STATE(b), and, additionally, y has been removed and x added to σ. The **forward canonical set of equations using maximum concurrency strategy** and the **forward complete canonical set of equations** for SMA are identical in this case:

$$x(\sigma) = \quad A1(a,b) \circ x(\sigma_{+b}^{-a})$$
$$\uplus \ (A1(a,e) \ \uplus \ A2(a,e)) \circ x(\sigma_{+e}^{-a})$$

$$\quad\quad \uplus\ (A1(a,c)\ \uplus A3(a,c))\circ x(\sigma_{+c}^{-a})$$
$$x(\sigma_{+b}^{-a}) = \ (A1(b,f)\ \uplus A2(b,f))\circ x(\sigma_{+f+(STEPS=2)}^{-a-(STEPS\neq2)})$$
$$x(\sigma_{+e}^{-a}) = \ (A1(e,c)\ \uplus A3(e,c))\circ x(\sigma_{+c+(STEPS=2)}^{-a-(STEPS\neq2)})$$
$$\quad\quad \uplus\ (A1(e,h)\ \uplus A3(e,h))\circ x(\sigma_{+h+(STEPS=2)}^{-a-(STEPS\neq2)})$$
$$x(\sigma_{+c}^{-a}) = \ (A1(c,f)\ \uplus A2(c,f))\circ x(\sigma_{+f+(STEPS=2)}^{-a-(STEPS\neq2)})$$
$$\quad\quad \uplus\ (A1(c,h)\ \uplus A2(c,h)\ \uplus A3(c,h))\circ x(\sigma_{+h+(STEPS=2)}^{-a-(STEPS\neq2)})$$
$$x(\sigma_{+f+(STEPS=2)}^{-a-(STEPS\neq2)}) = \ M1(A)\circ(M2(A_x,A_y)\ \uplus M3(A_x,A_y)\ \uplus M4(A_x))$$
$$\quad\quad \circ\ M5(A)\circ X(\sigma_{+f}^{-a})$$
$$x(\sigma_{+h+(STEPS=2)}^{-a-(STEPS\neq2)}) = \ M1(A)\circ(M2(A_x,A_y)\ \uplus M3(A_x,A_y)\ \uplus M4(A_x))$$
$$\quad\quad \circ\ M5(A)\circ X(\sigma_{+h}^{-a})$$
$$x(\sigma_{+c+(STEPS=2)}^{-a-(STEPS\neq2)}) = \ M1(A)\circ(M2(A_x,A_y)\ \uplus M3(A_x,A_y)\ \uplus M4(A_x))$$
$$\quad\quad \circ\ M5(A)\circ X(\sigma_{+c}^{-a})$$
$$X(\sigma_{+f}^{-a}) = \ t_1^1\circ X(\sigma_{+d}^{-a})$$
$$X(\sigma_{+h}^{-a}) = \ t_2^1\circ X(\sigma_{+d}^{-a})$$
$$X(\sigma_{+c}^{-a}) = \ t_3^1\circ X(\sigma_{+d}^{-a})$$
$$X(\sigma_{+d}^{-a}) = \ \varepsilon$$

Three temporary transits t_1^1, t_1^1 and t_3^1 have been added to terminate computations and to estimate the cost of remaining solutions.

The costs of temporary transits can be estimated, for instance, as at least equal to the minimum of costs of outgoing links for a given node, i.e. $t_1^1.cost=1$, $t_1^1.cost=1$, and $t_3^1.cost=1$.

i=1 EXAMINE phase:

The least fixed point to the set of equations from the select phase:

$$X(\sigma) = \ A1(a,b)\circ(A1(b,f)\ \uplus A2(b,f))\circ M1(A)\circ$$
$$\quad\quad \circ\ (M2(A_x,A_y)\ \uplus M3(A_x,A_y)\ \uplus M4(A_x))\circ M5(A)\circ t_1^1$$
$$\quad \uplus\ (A1(a,c)\ \uplus A3(a,c))\circ(A1(c,f)\ \uplus A2(c,f))\circ M1(A)\circ$$
$$\quad\quad \circ\ (M2(A_x,A_y)\ \uplus M3(A_x,A_y)\ \uplus M4(A_x))\circ M5(A)\circ t_1^1$$
$$\quad \uplus\ (A1(a,c)\ \uplus A3(a,c))\circ(A1(c,h)\ \uplus A2(c,h)\ \uplus A3(c,h))\circ M1(A)\circ$$
$$\quad\quad \circ\ (M2(A_x,A_y)\ \uplus M3(A_x,A_y)\ \uplus M4(A_x))\circ M5(A)\circ\circ t_2^1$$
$$\quad \uplus\ (A1(a,e)\ \uplus A2(a,e))\circ(A1(e,h)\ \uplus A3(e,h))\circ M1(A)\circ$$
$$\quad\quad \circ\ (M2(A_x,A_y)\ \uplus M3(A_x,A_y)\ \uplus M4(A_x))\circ M5(A)\circ t_2^1$$
$$\quad \uplus\ (A1(a,e)\ \uplus A2(a,e))\circ(A1(e,c)\ \uplus A3(e,c))\circ M1(A)\circ$$
$$\quad\quad \circ\ (M2(A_x,A_y)\ \uplus M3(A_x,A_y)\ \uplus M4(A_x))\circ M5(A)\circ t_3^1$$

The optimal solution, assuming that cost(a $\uplus$ b)=min(cost(a),cost(b)), and cost(a $\circ$ b)=cost(a) + cost(b), are any actions leading through nodes e and c (fitness of all actions is equal to 0, and their costs is predetermined by the costs of links). Let's assume that these are actions A2(a,e) and A3(c,e).

From the set of Non-Deterministic choices, the best solution was: $X(\sigma) =$ A2(a,e) $\circ$ A3(e,c) $\circ$ M1(A) $\circ$ (M2(A_x,A_y) $\uplus$ M3(A_x,A_y) $\uplus$ M4(A_x)) $\circ$ M5(A) $\circ$ t_3^1 . Hence for the first optimization step the best choice is to choose the sequence of nodes e, c as the next hop to destination d. The temporary transit t_3^1 is removed and the remaining part is directed for execution.

i=1 EXECUTE phase:

A2(a,e) and A3(e,c) are executed, and next bucket brigade modification M1 will increase fitness of optimal actions A2 and A3 ($\gamma_2 = \gamma_3 = -1$) and will decrease fitness of non-optimal action A1 tried in routing, but not chosen for execution ($\gamma_1 = 1$). M2, M3 and M4 will choose the fittest actions A2 and A3 for modifications to produce new offsprings. Crossover M3 of A2 and A3 does not lead to new actions, and mutation M4 leads to action which does not have chance to fire. Thus the inference engine selects specialization for execution. The learning phase gave rise to the creation of transit A6 (with fitness of its parents $\gamma_6 = \gamma_2 = \gamma_3 = -1$) which resulted from specialization of A2 and A3.

ACTION A6(X,Z)
  pre : STATE(X) $\wedge$ LINK(X,Z) $\wedge$ (STEPS $\neq$ 2)
      $\wedge$ LINK-LESS-THAN(3,X,Z) $\wedge$ NEIGHBOR-GREATER-THAN(1,Z)
  act : INCLUDE-LINK(X,Z) $\wedge$ (STEPS++)
  post: STATE(Z) $\wedge$ LINK(X,Z) $\wedge$ ((STEPS $\neq$ 2) $\vee$ (STEPS = 2))
      $\wedge$ LINK-LESS-THAN(3,X,Z) $\wedge$ NEIGHBOR-GREATER-THAN(1,Z)
  cost: link(X,Z).cost + γ_6

Because state σ_{+c}^{-a} does not match the goal, the inference engine will return to the select phase. The second phase i=2 of the expansion, for $k = 5$ is as follows:

i=2 SELECT phase:

$x(\sigma_{+c}^{-a}) =$ (A1(c,f) $\uplus$ A2(c,f)) $\circ$ $x(\sigma_{+f}^{-a})$
$\qquad \uplus$ (A1(c,h) $\uplus$ A2(c,h) $\uplus$ A3(c,h) $\uplus$ A6(c,h)) $\circ$ $x(\sigma_{+h}^{-a})$
$x(\sigma_{+f}^{-a}) =$ (A1(f,d) $\uplus$ A2(f,d)) $\circ$ $x(\sigma_{+d+(STEPS=2)}^{-a-(STEPS\neq2)})$
$x(\sigma_{+h}^{-a}) =$ (A1(h,f) $\uplus$ A2(h,f)) $\circ$ $x(\sigma_{+f+(STEPS=2)}^{-a-(STEPS\neq2)})$
$\qquad \uplus$ (A1(h,g) $\uplus$ A2(h,g)) $\circ$ $x(\sigma_{+g+(STEPS=2)}^{-a-(STEPS\neq2)})$
$x(\sigma_{+d+(STEPS=2)}^{-a-(STEPS\neq2)}) =$ M1(A) $\circ$ (M2(A_x,A_y) $\uplus$ M3(A_x,A_y) $\uplus$ M4(A_x))
$\qquad \circ$ M5(A) $\circ$ X(σ_{+d}^{-a})
$x(\sigma_{+f+(STEPS=2)}^{-a-(STEPS\neq2)}) =$ M1(A) $\circ$ (M2(A_x,A_y) $\uplus$ M3(A_x,A_y) $\uplus$ M4(A_x))
$\qquad \circ$ M5(A) $\circ$ X(σ_{+f}^{-a})
$x(\sigma_{+h+(STEPS=2)}^{-a-(STEPS\neq2)}) =$ M1(A) $\circ$ (M2(A_x,A_y) $\uplus$ M3(A_x,A_y) $\uplus$ M4(A_x))
$\qquad \circ$ M5(A) $\circ$ X(σ_{+h}^{-a})
$X(\sigma_{+f}^{-a}) = t_1^2 \circ X(\sigma_{+d}^{-a})$
$X(\sigma_{+h}^{-a}) = t_2^2 \circ X(\sigma_{+d}^{-a})$
$X(\sigma_{+d}^{-a}) = \varepsilon$

The estimated cost of two temporary transits t_1^2 and t_2^2 is 1.

i=2 EXAMINE phase:

From the set of Non-Deterministic choices, the best solutions was: $X(\sigma_{+c}^{-a}) = $
A2(c,f) ∘ A2(f,d) ∘ M1(A) ∘ (M2(A_x,A_y) ∪ M3(A_x,A_y) ∪ M4(A_x)) ∘ M5(A)

For the second optimization step the best choice is to choose the sequence of nodes f,d as the next hop to destination d.

i=2 EXECUTE phase:

A2(c,f) and A2(f,d) are executed. M1 updates fitness of actions: $\gamma_1 = -2$, $\gamma_2 = 2$, $\gamma_3 = \gamma_4 = \gamma_6 = 0$. The inference engine picks up actions A4 and A6 for modification (high fitness) and performs on them crossover M3 (it has a nondeterministic choice between M2, M3 and M4). M5 is iddle because transits are still sufficiently "fit". The learning phase gave rise to the creation of transits A7 which resulted from crossover of actions A4 and A6 (the second created transit is the "old" A3).

ACTION A7(X,Z)
    pre : STATE(X) ∧ LINK(X,Z) ∧ (STEPS ≠ 2)
           ∧ LINK-LESS-THAN(2,X,Z) ∧ PATH-LESS-THAN(3,Z)
    act : INCLUDE-LINK(X,Z) ∧ (STEPS++)
    post: STATE(Z) ∧ LINK(X,Z) ∧ ((STEPS ≠ 2) ∨ (STEPS=2))
           ∧ LINK-LESS-THAN(2,X,Z) ∧ PATH-LESS-THAN(3,Z)
    cost: link(X,Z).cost + γ_7

The terminal state d has been reached thus stop to the inference engine $k\Omega$-Procedure. The optimal sequence is a,e,c,f,d, executed by actions A2(a,e), A3(e,c), A2(c,f), and A2(f,d).

By looking at this result it can be seen that after the first phase of the expansion, the learning process created the action *A6* which resulted from the combination of actions *A2* and *A3*. This new action was tried in the second phase, and then the learning process gave rise to further exploitation of this action when action *A7* was created from the crossover operation of actions *A6* and *A4*.

The resulting routes proved to be loop free and optimal mainly because if the method of operation of the inference engine. This is mainly because:

- The SMA Inference Engine examines all possible paths within k steps and selects the path of cheapest cost. Hence the algorithm always provide the most optimal path available.
- The algorithm always provide loop free paths and this is also because of the inference engine optimization techniques. If ever the expansion of the transits lead to loops in the path, then this path would obviously have a larger cost than the solution without the loop. Since the inference engine examines possible solutions and selects the cheapest, the solution that contains loops would obviously not be selected.

6 Conclusions

This paper identified the strengths and weaknesses of some of the existing methods of finding optimal paths in a network. Existing routing algorithms work well, except when faced with the problem of multiple and rapid link failure or when the network itself is too unstable for proper decisions to be made effectively. When such changes occur in the network there was therefore the need to find alternative paths through which information can be routed so that packets on the network does not experience any major delay. In the case of very large networks, with many links there are various possibilities of finding an alternative paths and therefore some method has to be devised where intelligent decisions are to be made as to which of the alternative paths to explore. In order to make such intelligent decisions there was a need to implement some sort of machine learning techniques where the system was able to learn good routing techniques based on its past experience. Such a system therefore required the capability of discovering new heuristics and the testing and exploiting such heuristics in such a way that the performance of the system does not suffer.

The SMA inference engine has been shown to be very effective in incorporating various forms of machine learning techniques, as well as the techniques of evolutionary computing to model a system that has similar capabilities of the classifier system proposed by Holland [14, 12]. It seems that CSA is more general than genetic algorithms/evolutionary programming, neural networks, or conventional symbolic inference techniques [12, 8], because CSA allows relatively easily to express and combine all these approaches as the special cases of more general CSA modifications. The CSA optimization mechanism can be used to increase fitness of genetic algorithms, or to minimize error of classification in neural nets, or to provide fault-tolerance. However, we are only at the initial stage to investigate and explore CSA capabilities. In the case of computer network routing, the inherent parallelism and nondeterminism incorporated in the CSA provides multiple possibilities of alternative routes, with each possibility being selected by the cheapest routing heuristic, thus providing potential improvement in the performance of the network routing. The "routing" SMA was tested and proved capable of learning using a simple network mesh topology [20] and the SEMAL programming language [11]. However, the results are preliminary yet. A current version of SEMAL [2, 16] does not have implemented transit activities thus actvities of actions and modifications have been simulated "by hand" only. To obtain meaningful results for more complex "real" networks, additional computing requirements are necessary so that it would be possible to explore and exploit a wide variety of possible techniques at different levels of learning, and to compare the performance of the proposed method with known routing algorithms.

References

1. Bertsekas D., Gallager R., *Data Networks* Englewood Cliffs, NJ: Prentice-Hall, 1992.

2. Blondon R., *Towards an Implementation of SEMAL: Building the set of Equations*, Honours Thesis, Jodrey School of Computer Science, Acadia University, August 1993.

3. Cheng C., Cimet I. A., and Kumar S. P. R., "A Protocol to Maintain a Minimum Spanning Tree in a Dynamic Topology," *ACM SIGCOM Symp. Comm. Arch. and Protocols*, pp 330 - 338, Stanford, CA, Aug. 1998.

4. Cheng C., Riley R., Kumar S.P.R., Garcia-Luna-Aceves J.J., "A Loop-Free Extended Bellman-Ford Routing Protocol without bouncing effect" *SIGCOMM '89 Symposium* Communications Architecture and Protocols, Austin Texas, Sept. 1989.

5. Cohen P. R., Feigenbaum E. A., The Handbook of Artificial Intelligence : Volume 3, William Kaufmann Inc., 1982

6. Cimet I. A., Cheng C., and Kumar P. R., "On Design of Resilient Protocols for Spanning Tree Problems," *IEEE Int'l Conf. on Distributed Computing*, June 1989.

7. Eberbach E.,"Self-Modifiable Algorithms and their Applications," *Research Note RN/88/27*, Department of Computer Science, University College of London, (June 1988).

8. Eberbach E., "CSA: In the Direction of Greater Representational Power for Neuro-computing," *Journal of Parallel and Distributed Computing*, Vol. 22, No. 1., 1994, 107-112.

9. Eberbach E., "Selected Aspects of the Calculus of Self-Modifiable Algorithms Theory," *Lecture Notes on Computer Science* 468, Springer-Verlag, 1990, 34-43.

10. Eberbach E., "Fixed Point Semantics for sets of Equations over SMA-Net", Proceedings of the Second International Conference on Fixed Point Theory and Applications, Halifax, Canada, 1991, 97-111.

11. Eberbach E., "The Design and Specification of SEMAL - A Cost Language Based on the Calculus of Self-Modifiable Algorithms," *Proc. of the 5th Int'l Conf. on Software Engineering and Knowledge Engineering* SEKE'93, San Francisco, California, 1993, pp. 166-173.

12. Eberbach E., "Neural Networks and Adaptive Expert Systems in the CSA Approach," *International Journal of Intelligent Systems*, Vol. 8, No. 4, 1993, pp. 569-602.

13. Fogel L. J., Owens, A. J., Walsh M. J., *Artificial Intelligence through Simulated Evolution* , New York: Wiley 1966.

14. Holland, J. H., *Adaptation in natural and Artificial Systems*, Ann Arbor, MI: University of Michigan Press. 1975.

15. Holland J. H., "Escaping Brittleness : The Possibilities of General-Purpose Learning Algorithms Applied to Parallel Rule-Based Systems," in Michalski et al, *Machine Learning. An Artificial Intelligence Approach*, vol. 2, Morgan-Kaufmann, 1986, Chapter 20.

16. Horree S., *Design and Implementation of a Version of the SEMAL Interpreter*, Honours Thesis, Jodrey School of Computer Science, Acadia University, April 1994.

17. Jaff J.M., Moss F. H., "A Responsive Distributed Routing Algorithm for Computer Networks," *IEEE Transactions on Communications*. vol. COM-30, NO. 7, July 1982.

18. Merlin P. M., Segall A., "A Failsafe Distributed Routing Protocol," *IEEE Transactions on Communications*. vol. COM-27, NO. 9, September 1979.

19. Ramachandran G., *Implementation of Concurrent Algorithms in OCCAM on Transputers and their Preliminary Specification in SEMAL*, Honours Thesis, Jodrey School of Computer Science, Acadia University, April 1993.

20. Seunarine D., *The CSA-Based Evolutionary Computer Network Routing Algorithms*, Master Thesis, Jodrey School of Computer Science, Acadia University, April 1994.

21. Tan V.K., Eberbach E., "A CSA-Based Robot Plan Generator," *Proc. of the 21st Annual ACM Computer Science Conference CSC'93*, Indianapolis, Indiana, (1993), 115-122.

22. Veerayah K., *The CSA-Based Distributed Fault-Tolerant Systems*, Honours Thesis, Jodrey School of Computer Science, Acadia University, April 1993.

Evolving Robot Strategy for Open ended Game

Tomonori Sugiyama and Takashi Kido and Masakazu Nakanishi

Department of Mathematics, Keio University, 3-14-1, Kanagawa, 223, Japan

Abstract. A good entertaiment must be interesting. This paper asserts that interesting games have unpredictability. Recent technologies in artificial life give us new possibility of unpredictability, such as the evolution of strategies of opponents. This paper seeks this possibility in a robot battle game using genetic algorithm for the evolution of strategies. We made a robot battle game called X-Window Robot Battle (XRB), a fighting game of two robots. Each robot's strategy is given by assembly language codes, and these codes are created by a user or a computer. We use genetic algorithms (GA) to evolve a robot's codes made by computer and make it possible for the robot to acquire a useful strategy without being explicitly programmed. The change of strategy of an opponent makes the game more unpredictable and interesting, thus we can enjoy the game. We believe that our attempt contributes to the entertainment industries.

1 Introduction

A good entertainment must be interesting. What is an interesting entertainment? What makes us feel attractive? We consider it in games. Some games such as Chess, Go and Shogi have been enjoyed by many people for a long time, thus it can be concluded that they are interesting. It seems that these games are interesting because of their complexity. The complexity comes from not only the vastness of the game's search space but also unpredictness of the opponent's behavior[1].

Some games use random numbers for its parameters and they can make us feel unpredictable. However, once we know how a game uses their parameters, it will be eventually predictable, the game becomes uninteresting, and the game ends its role as an entertainment.

Recent technologies[5][6] in artificial life give us another possibility of unpredictability, such as the evolution of strategies of virtual opponents. Evolution of virtual opponents in a game makes it open-ended.

We believed that really interesting games are 'open-ended games'. The purpose of this paper is to provide a sample of 'open-ended games' using AI and A-LIFE technologies.

We made a robot battle game called an X-Window Robot Battle (XRB), a fighting game of two robots. Each robot's strategy is given by assembly language codes, and the codes are written by a user or produced by a computer. We use genetic algorithms (GA)[2][3] to evolve a robot's codes made by computer and make it possible for the robot to acquire a useful strategy without being explicitly programmed. The change in a strategy of opponents makes the game more unpredictable and interesting, thus, most importantly, enjoyable.

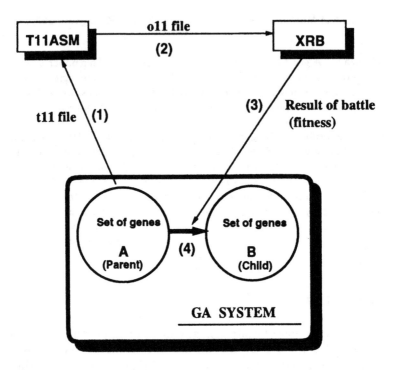

Fig. 1. System overview

2 System overview

Figure 1 shows the composition of our system. The system has three parts, T11ASM, XRB, and GA system. T11ASM is an assembler to make an object program for XRB. Programming language for XRB is based on an assembly language from PDP-11. We call this assembly language T-11. And each operation of the robot is provided to a player as a subroutine. Table 1 shows an example of a program. XRB is a battle game of two robots. All robots have the same specifications, therefore only the operating program makes it possible to determine the ability of the robot. GA system produces the robot's action algorithm using percentages of wins in XRB's battle.

(1) T11ASM accepts T11 source program that produced by genes in the GA system. (2) XRB accepts object files made by T11ASM. (3) Finishing all the

battles with other robots, the GA system gets the result as a fitness function. (4) And using with these data, the GA system creates strategies of robots using GA techniques.

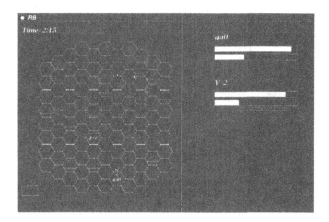

Fig. 2. XRB

3   Instruction of XRB

This section gives an overview of a game system which we have implemented. We have one robot in the game field. The robot cannot work in any way by itself, therefore we have to prepare a program to make the robot defeat an opponent in the game field. A game is over when damage of one of the players reaches 100 %, or battle time is exhausted of the later less damage player wins. Battle field can be classified into some types. The simplest type is a no hazard one. There are nothing other than the opponent and the player's robot itself. Other types are hazard models. In the field, various kinds of hazards are located randomly.

Such games have been presented many times untill before this paper, and most of them have some kind of programming environment in their system. These systems are, however, accepted only by programmers because they have to make efficient programs to defeat the opponents.

Here, we introduce rules of the game. First of all, our game has enough visual user interface which can display the moves of the robots. This interface is very important for this kind of game, because unpredictable move of the robot is very attractive for us.

Figure 2 shows the hex-map of XRB. The battle field is within the area of hexagons where robots can move to six directions. Each robot is drawn with a circle with a tag (e.g. ga0, Y-2) that shows the name of the robot. Other objects

```
;;    example robot T-11 program

;;    subroutine entries
      FORWARD = 100002
      SHOOT   = 100016
      CHKSLF  = 100026
      DIRSRC  = 100032
      DISSRC  = 100036
      FORD    = 1
      FDRT    = 4
      FDLT    = 2000
      .=0

;; Start program
START:  JSR       PC,SCAN
        BR        START

;; Scan Enemy
SCAN:   CLR       R0
        JSR       PC,DIRSRC
        BIT       #FORD,R0
        BNE       BOMB              ;Check Front
        BIT       #FDRT,R0
        BNE       2$                ;Check Front Right
        BIT       #FDLT,R0
        BNE       3$                ;Check Front Left
        MOV       #1,R0
        JMP       FORWARD           ;Go ahead
2$:     MOV       #20,R1            ;Shot Front Right
        BR        FIRE
3$:     MOV       #40,R1            ;Shot Front Left
        BR        FIRE

;; Shot Canon
FIRE:   CLR       R0
        JSR       PC,DISSRC
        CMP       #1,R0             ;distance=1 ?
        BEQ       PUNCH
        MOV       R1,R0
        JSR       PC,SHOOT
        RTS       PC

;; Bomb !!
BOMB:   MOV       #1,R0             ;Check Bomb rest
        JSR       PC,CHKSLF
        TST       R0
        BEQ       FIRE
        CLR       R0
        JSR       PC,DISSRC
        CMP       #2,R0             ;distance<4 ?
        BPL       FIRE
        MOV       #2,R0             ;Bomb
        JSR       PC,SHOOT
        BR        FIRE

;; Punch
PUNCH:  MOV       #3,R0             ;Punch
        JSR       PC,SHOOT
        RTS       PC
        .END      START
```

Table 1. Example T-11 Program for XRB

in the field are hazards — mountains, ponds, etc. For each robots, the status is shown as graphs on the right side in the screen. The upper one means the damage of the robot, and the lower one is the energy left. The square just below is the number of available weapons; upper one are canon bullets, lower one are bombs. Finally, the time is counting down at the left-top of screen.

Some operations for a robot such as "forward", "back" and "search opponents in the any direction", "check one's energy" are provided . These operations are sets of assembly codes. Each operation of the robot is provided to a player as a subroutine. All the operations of the robot are listed in Table 3.

4 Program synthesis by GA

```
(1 go ahead
    (2 Enemy before?
        (3 distance=1? (4 punch (goto 1))
                       (5 shot (goto 1)))
        (goto 1)))
```

Fig. 3. S-expression

Major interest of this kind of game is "how your programmed robot works and defeats the opponent". However it can also give us large interests just watching how robots of the various concepts work, especially, the robot that is programmed by a computer. From this point of view, we have done some experiments to examine how effective GA works when it is applied to the game. The strategies of robots are created using GA techniques. The basic idea of this technique is almost same as a genetic programming approach [4]. The robot based on a weak strategy are defeated by others, while the one based on a strong strategy defeats others and make his child. Hopefully the robot with enough strong strategy can defeat its opponent, even if the opponent is programmed by a human.

It is possible to use tree structures to represent a robot program. We show a simple example of a robot action algorithm (Figure 4) and its S-expression (Figure 3).

In our model, a robot is controlled by T-11 assembly codes such as MOV, TST, etc. It is possible to use this codes as genes of GA directly, however it has some problems. First, randomly generated codes have a risk of containing errors as machine codes. Second, codes generated by GA operator also have a risk of containing errors as machine codes. For this reason, GA system produces error-free S-expression like representations. Then the S form is converted to T-11 assembly codes. Table 2 shows an example of S-expression converted by GA system.

```
;;      example robot algorithm by GA
        FORWARD = 100002
        TURNR   = 100006
        TURNL   = 100012
        SHOOT   = 100016
        BACK    = 100022
        CHKSLF  = 100026
        DIRSRC  = 100032
        DISSRC  = 100036
        DISPRZ  = 100042
        .=0
START:
L1:
        ;(1)Go ahead
        MOV     #1,R0
        JSR     PC,FORWARD
        JMP     L2
;
L2:
        ;(101)Enemy before?
        CLR     R0
        JSR     PC,DIRSRC
        BIT     #1,R0
        BEQ     0$
        JMP     L3
0$:     JMP     L1
;
L3:
        ;(113)Distance=1 ?
        CLR     R0
        JSR     PC,DISSRC
        BIT     #1,R0
        BEQ     0$
        JMP     L4
0$:     JMP     L5
;
L4:
        ;(15)Punch
        MOV     #3,R0
        JSR     PC,SHOOT
        JMP     L1
;
L5:
        ;(12)Shoot
        CLR     R0
        JSR     PC,SHOOT
        JMP     L1
;
        .END START
```

Table 2. Example Program produced by GA

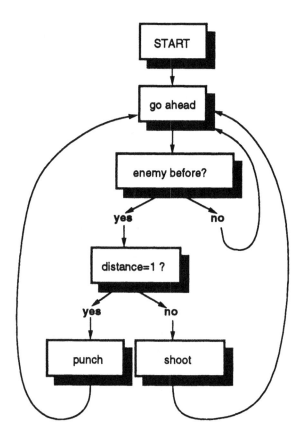

Fig. 4. Representation of robot action

We use three conventional operations of GA to synthesize the program : namely selection, crossover and mutation. Clearly, the robot that can win in high probabilities should be given the high fitness. Each robot produced by the GA system fights against all of robots written by human. Therefore, fitness value of GA is a winning rate in the battles.

In the steps of selection, parents are chosen in the ratio of the fitness value of one robot to the sum of the fitness values of all the robots generated by GA.

Crossover operations on tree structures are done by exchanging subtrees. Subtrees are selected at random. Figure 5 shows the crossover action of this system.

We take three kinds of mutation: 1) to change an operation, 2) to change a jump label and 3) to replace a branch to a newly created one.

By using the tree structure, we can not only represent the robot program, but also find good building blocks efficiently. The inherited building blocks are important parts of the tree structure, because they have all the strategies of the parent programs after the selected node, except for the goto actions.

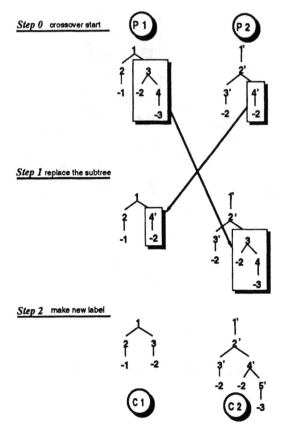

Fig. 5. Crossover operation on tree structures

5 Experimental Results

5.1 Experiment 1: Adaptation to different opponents

The purpose of this system is to produce strong program. The strong program means that the robot controled by this program can defeat the enemys as many times as possible. We believe that evolutionary process can produce such kind of programs.

A set consists of 5 robots made by human. We prepared 3 sets of 5 robots each of which has a different concept with one another. We tried 100 trials with

Forward	move forward by specified steps
Turn Right	turn right by specified steps
Turn Left	turn right by specified steps
Back	move backward by specified steps
Shoot	fire specified weapon(bomb and canon) or punch
Check	return the status of weapon, damage and energy
Direction Search	search an opponent or an object at all direction (no concern about distance).
Distance Search	search an opponent or an object at specified direction and distance.
Random	return a random value.

Table 3. Operations of Robot

100 generations of 40 robots. The following result is a score of winning average of 100 trials of 100th generations.

Set	win (%)	lost (%)	draw (%)
Set 1	63	3	34
Set 2	51	20	29
Set 3	53	8	39

We conclude that every set generates effective robots.

5.2 Experiment 2: Effective of Crossover and Mutation

The next generation is made from the current generations with the operation 'crossover' and 'mutation'. In order to examine the efficiency of each operation, we use three kinds of programs generated with both crossover and mutation, with only crossover, and with only mutation.

Each experiments do 100 trials with 100 generations of 40 robots. The following is the result of the same experiment done in experiment 1.

operation	win (%)	lost (%)	draw (%)
Both	53	41	6
Only Crossover	68	20	12
Only Mutation	21	63	16

We try 3 kinds of experiments that are both with crossover and mutation, with only crossover, and with only mutation. We conclude that crossover is very effective, however on the contrary, it seems mutation has bad effects on the evolution. This is because effective mutation hardly occurs.

5.3 Experiment 3: Influences from uncertain factors

We also examine the reactions from the uncertain factors which are not the opponents themselves. In this paper, we use 2 factors, one determines whether we set hazard or not, and another determines whether we change a starting position or not. The influences from the former factor would be small, the latter's would be large.

We use 4 type factors in this experiment. The simplest type is non-hazards and non-random set up, the second type has hazards in battle field, and the third type is to set starting point at random. The last type is to use both of factors, random hazards and random set-up.

factors	win (%)	lost (%)	draw (%)
Normal	55	13	32
Random Hazards	19	41	40
Random Set-up	11	70	19
Both	14	52	34

We cannot get good result in this experiment. Many games end in a draw, because the two robots couldn't encounter so often during the game. And in this experiment, this system cannot adapt robots to the uncertain factors in XRB. This reason is that many kinds of situations occurred by these factors. For example, Opponent is hided behind mountains or located too far by random set-up.

6 Conclusion

We provide XRB, a fighting game of two robots. Each robot's strategy is given by assembly language codes, and those codes are created by a user or a computer. We use GA to evolve a robot's codes made by computer and make it possible for the robot to acquire a useful strategy without being explicitly programmed. The robot evolved by GA is competitive with the one by a user. In fact, in our experiment, winning average is over 50 % all of sets, and hardly lost. In this paper crossover is very efficiently, however mutation is not. Mutation operation destroies effective building blocks. Crossover operation makes progress in establishing a better strategy.

So far, our current system cannot adapt robots to the uncertain factors in XRB. This is our future work.

The changes in a strategy of opponent make the game more unpredictable and interesting, and we have enjoyed. In fact, the player of XRB is very exciting especially when his program is defeated by GA. We believe that our attempt contributes to the entertainment industry.

References

1. Tomonori Sugiyama, Takashi Kido, Mutsuhiro Yonezu, Junya Tsutsumi, Fumihiko Yamaguchi, Masaaki, Hayashi, Masakazu Nakanishi, *Program Synthesis for Robot Wor using Genetic Algorithm.* Workshop on Artificial Intelligence and Artificial Life for Entertainment 1994.
2. David E. Goldberg, *Genetic Algorithms in Search, Optimization Machine Learning.* Addison-Wesley, Reading, Mass, 1989.
3. J. H. Holland, *Adaptation in natural and artificial systems* The University of Michigan Press, 1975.
4. John R. Koza, *Genetic programming* The MIT Press, 1992.
5. R. Axelrod and D. Dion. The Further Evolution of Cooperation, Science, vol. 242, pages 1385-1390, December 1988.
6. K. Lindgren. *Evolutionary Phenomena in Simple Dynamics* In C. Langton, C. Taylor, J.D. Farmer, and S. Rasmussen, editors, *Artificial Life II.* SFI Studies in the Science of Complexity, 1991.

An Evolutionary Approach to Adaptive Model-Building *

Zhengjun Pan   Lishan Kang   Jun He   Yong Liu

Institute of Software Engineering, Wuhan University

Wuhan 430072, P.R.China

Abstract

Learning an input-output mapping from a set of examples can be regarded as a problem of model-building. From this point of view, by a hierarchical representation scheme for models, an evolutionary computational approach to model-building problems is proposed in this paper, which is based on the ideas from the evolution programs and the genetic programmings. The computer experiments indicate that it is surprisingly effective in searching the optimal model for model-building problems.

1. Learning As Model-Building

The problem of learning a mapping between an input and an output space is equivalent to the problem of estimating the system that transforms inputs into outputs given a set of examples of input-output pairs [8]. It is also equivalent to the problem of building a model to represent the system , which is called a model-building problem. Several classical approaches to this problem are data fitting, regression analysis and approximation theory. These techniques usually deal with the problem of fitting or approximating or interpolating a set of input-output pairs $\{(x_i, y_i), x_i \in X, y_i \in Y\}$ by a function (sometimes we call it a model) $f(x, w)$ having a fixed number of parameters w belonging to some set W (x and w are real vectors $x = (x^1, x^2, \cdots, x^n) \in X, w = (w^1, w^2, \cdots, w^m) \in W, f(x, w) \in Y$). For a choice of a specific function f, the problem is then to find the set of parameters w that provides the least discrepancy of the expected values $\{f(x_i, w)\}$ from the observed values $\{y_i\}$. Needless to say, it is very important to choose an appropriate model $f(x, w)$ as well as possible, for the reason that it would be little point in trying to optimize the parameters w, if the chosen model $f(x, w)$ could only give a very poor representation of $\{(x_i, y_i)\}$, even f has optimal parameter values. However, it is usually an arduous work to choose an appropriate model $f(x, w)$ as well as to find

*This work was supported in part by National Natural Science Foundation of China and National 863 High Technology Project of China.

the optimal values of the parameters w, especially when we do not grasp well the background of the original problem[5].

The newly developed artificial neural network techniques also provide an alternative approach to the model-building problem. In order to make an artificial neural network model to perform well, the network must have an appropriate structure, which includes the topology of nodes and links as well as the weight values. However, it is also very difficult to design an appropriate artificial neural network structure[6][10] for a specific problem.

Because of the limitations of the classical techniques for model-building problems, we propose a new method based on evolutionary ideas. In this method we start with a population of individuals. Each individual represents a potential solution to the problem. Through some selection (selecting more fit individuals) and recombination (forming new models) operators which are based on the fitness of the individuals, a new population is formed. The new generation usually has some more fit models. After some number of generations, the best individual is the model hopefully representing the system.

A. Model-Building Problem

Many problems in various applied fields can be formulated as a model-building problem or , equivalently, as a problem of learning from examples. To build a model means to provide a convenient conceptual representation of the observed system from examples. It is on the model which we use to think about the true situation that interpretations, predictions, and decisions would be based.

Successful model fitting is a matter of ensuring that there are sufficiently data to achieve the required level of accuracy. Usually the quality of a model is measured by an objective function , which is related to the discrepancy of the observed values $\{y_i\}$ from the expected values $\{f(x_i)\}$ and is normally induced by a distance function ρ. Then the model-building problem can be stated formally as:

Model-Building Problem: Given a set of input-output pairs $\{(x_i, y_i); x_i \in X, y_i \in Y, i \in I\}$, where X, Y are subsets of real finite dimensional spaces respectively, I is an index set which can be a finite or an infinite set . If F is a subset of $C(X)$ (the set of all continuous functions defined on X) . The model-building problem is to determine the function $f^* \in F$ such that

$$\rho(\{f^*(x_i)\}, \{y_i\}) \leq \rho(\{f(x_i)\}, \{y_i\}) \tag{1}$$

for all $f \in F$, or

$$\rho(\{f^*(x_i)\}, \{y_i\}) \leq \epsilon \tag{1}'$$

where ϵ is a small given number.

A solution to this problem, if it exists, is said to be a best model. The existence of a best model depends ultimately on the function class F to whom f belongs.

The inequality (1) or (1)' illustrates that a model-building problem is actually an optimization problem , in which we are required to find a function (from class F) to minimize the objective function

$$q(f) = \rho(\{f(x_i)\}, \{y_i\})$$

Without loss of generality, we next only consider the case that X and Y are both a subset of the real space R^1.

The model-building problem discussed above is in the case of without constraints. It is necessory to point out that the constrained case can be managed in a similar way.

B. Evolution Programs

Genetic algorithms (GAs) , first specified by John Holland in the early seventies[3], are becoming an important tool in optimization problems and machine learning. To solve a problem, a genetic algorithm maintains a population of potential solutions which are coded as fixed-length binary strings and probabilistically modifies the population by two basic genetic operators(selection and recombination), seeking a near-optimal solution to the problem.

From [1] we know that the binary coding offers the maximum number of schemata, but as demonstrated by Z. Michalewicz in [7], binary representations are not quite natural for many problems, and sometimes it may delay finding the optimal solution with the desired precision. Especially, when the solution space is quite large (or infinite) and high precision is required, the binary coding would require prohibitively long strings. This will prohibit GA to run efficiently.

Evolution programs (EPs) developed by Z. Michalewicz are also based on the principles of natural selection and natural genetics. The main difference from GAs is that EPs allow any data structure suitable for the problem together with the "genetic" operators approriate for this given data structure . This approach, unlike GAs, does not force us to represent the problem in such a way that solutions are specified by binary strings. It leaves the oringinal problem unchanged, only modifies the chromosome representations and applies specific "genetic" operators on them.

2. Representation

For the model-building problem, searching for an optimal model is carried out in a specific function space F (we call it a model family) , while F can be viewed as generated by some simpler functions through finite number of compositions and operations. Consequently, a quite natural representation of functions in F is to represent them by some simpler ones with operations, just like we represent elementary functions by the fundamental ones. From this point of view, we have

Definition 1   Let F be a function space, P be a finite subset of F, M be a finite set of mappings from either F to F or $F \times F$ to F, which are called unitary and binary operations on F respectively. If every function in F can be obtained from elements of P via finite number of operations belonging to M, then we say that P is a prototype set of F, M is an operation set of F, and denote $F = span(P, M)$.

For example, if $P = \{x, \alpha; x$ is the identity function, $\alpha \in R^1$ represents the constant function$\}$, $M = \{+, -, *, /\}$, then $span(P, M)$ is the set of all rational functions.

From definition 1, it is easy to show that if P, M are a prototype set and an operation set of F respectively, then every function f in F can be represented by a string of $P \cup M$. However, not every string of $P \cup M$ represents a function in F.

Definition 2 A string of $P \cup M$ representing $f \in F$ is called a string representation of f. If a substring of the string representation of f represents a function in F, the substring is called a legal substring of f.

For example, a string representation of the rational function $(x + 2)/(x^2 - 5)$ is (x 2 + x x * 5 - /) (the polish postfix notation) and (x x * 5 -) is a valid substring but (+ x x * 5) is not.

This particular string representation of $f \in F$ is corresponding to the polish postfix notation of f, hence it corresponds to a binary tree. These binary trees have the property that the elements of the terminal nodes belong to P and the others belong to M. It is obvious that every binary tree with these properties corresponds to a string representation of a function f (postorder traversal of the tree), and that every subtree corresponds a legal substring of f, and vice versa.

Definition 3 The binary tree corresponding to the string representation of f is called the (binary) tree representation of f.

In our experiments, we use binary tree representations to avoid checking the legality of substrings of string representations. This representation scheme is inspired to some extent from [4].

As we know, the conventional genetic algorithm is very effective for rapidly finding the general neighborhood of the optimal solution in a large search space, by using fixed-length binary strings. But the required precision, which is related to the representation scheme, is pre-determined. This limitation can prevent it from finding a highly precise solution in the long run. However, our representation is a hierarchical structure. It can vary adaptively as solving the problem. This makes it possible to search a large search space for the correct solution with an even-higher degree of precision.

3. Genetic Operators

Because we use "natural" data structures as the chromosome representations of potential solutions to a problem, the operators used in this case will be quite different from the traditional ones for the following reasons:

1) We deal with binary trees rather than binary strings.

2) The length of a string representation with which we represent a model is not fixed as in GAs. That is to say, any two binary trees in a population are typically of unequal size (different number of nodes and/or different depth) .

3) The allele (gene value) type in our case is not the boolean type. Thus the mutation operator is not just to change a 1 to a 0 (or vice versa) but more complex.

4) Some genetic operators are non-uniform, i.e., their action depends on the generation number.

These differences intensely affect the genetic operators, which can be seen from the following description of the operators.

Because of intuitive similarities, we only employ three operators: reproduction, crossover and mutation.

1) *Reproduction*: This is the standard roulette wheel reproduction operator[1]. It consists of two steps. First, a single model is selected from the population according to some selection method based on its fitness. Second, the selected individual is copied, without alteration, from the current population into the new one. As no

explicit local optimizing operator is used, we use rank selection strategy[2][9] to exaggerate the difference among similar fitness values for improving fine-tuning of the models. Rank selection is based on the rank of the fitness values (not on its numerical values) of models in the population. This will reduce the dominating effect of the comparatively high-fitness model in the selection procedure. Hence it can avoid the tendency of this model to dominate the population by establishing a predictable, limited amount of selection pressure on it. This will prevent premature convergence. The elitist strategy (preserving the best individual of the last generation in the new population) has also been used to get the best-so-far model.

2) *Crossover*: The crossover operation creates variation in the population by producing new models that consist of parts taken from each parent. The crossover operator starts with two parental models and produces two offspring for each two parents. The two parental models are also chosen from the population using the rank selection method. The operation begins by independently selecting one random point in each parent to be the crossover point. The crossover fragment of each parent is the subtree which has the crossover point as its root. Then the two offspring models are produced by deleting the crossover fragments of the two parental models and swapping them for each other. In our experiments, because of the limitation of computer memory and in order to avoid the model being too complex, we limit the representation length of the models by defining a maximum depth for the representation trees. Consequently, the chosen of the crossover points must obey the criterion that the two offspring models are within the maximum depth.

3) *Mutation*: This is quite different from the traditional mutation operator with respect to the number of the applicable alleles. In a traditional case, the applicable alleles in a locus (position of a gene) are just 0 and 1 , and the mutation operation is only to negate each bit with probability p_m (the probability of mutation) . But in our case, they are P (if the locus is a terminal point of the tree) and M (if the locus is an internal point of the tree) . Our mutation operator operates on one individual which is also selected by the rank selection method based on (normalized) fitness. It begins by selecting a point (called a mutation point) randomly within the model, then one of the following operations is performed with equal probability when the mutation point is an internal point of the tree:

- deleting the subtree which has the mutation point as its root (called a mutation subtree) and inserting a randomly generated subtree at that point.
- swapping the left subtree and the right subtree of the mutation subtree.
- simplifying the mutation subtree, i.e., replacing some parts with a value obtain from computation. For example, the string (3 5 +) will be replaced by 8 and (x x −) will be replaced by 0.

When the mutation point is a terminal point, we will replace it by an any other element in P if it is not a number , or multiply it by a random number in $(1-\delta, 1+\delta)$, where $\delta = (1 - \frac{t}{T})^2$, T is the maximum generation number and t is the current generation number.

4. Experimental Results

We implement the algorithm presented above on SUN SPARC 2 work stations. In our experiments, we take $P = \{x, \alpha; \alpha \in R^1\}$, $M = \{+, -, *, /, sin, cos, exp, ln\}$, then the model family $span(P, M)$ is a subset of all elementary functions. If $\{(x_i, y_i); x_i \in R, y_i \in R, i = 1, 2, \cdots, n\}$ is a set of given input-output pairs We define the objective function by

$$q(f) = \rho(\{f(x_i)\}, \{y_i\})$$
$$= \sum_{i=1}^{n} [f(x_i) - y_i]^2$$

If $q(f_i, t)$ is the objective function value of individual f_i in the population at generation t and $n_i(t)$ is the rank of $q(f_i, t)$ by sorting $\{q(f_i, t); i = 1, 2, \cdots, N\}$ from the maximum to the minimum, then the probability (i.e., the normalized fitness value) that individual f_i will be selected into the next generation of the population as a parent to be operated on by genetic operators is

$$r(f_i, t) = \frac{2 n_i(t)}{N(N+1)}$$

where N is the population size, i.e., the number of individuals in the population.

Then our problem is to find a model in $span(P, M)$ to minimize the objective function.

Because of the diversity of the functions during the computation, some of the functions generated may not be well-defined over the entire domain (e.g., division by zero or logrithm of zero at some point) or exceed the maximum number that can be represented on a computer (i.e., overflow), in these cases, we will use the penalty method to give a sufficiently large number to $q(f)$, then these functions will be abandoned hopefully after some generations.

A. A Test Problem

The set of data listed in Table 1 is taken from Example 4.7.7 of [11] , which is got by a business company to estimate the relation between the number of days (x) in professional training for salesmen and their bussiness achievement scores (y). In [11], through a logrithmic transformation, a model $f(x) = 31.71 * (1.363)^x$ is obtained by linear regression.

i	1	2	3	4	5	6	7	8	9	10
x_i	1	1	2	2	3	3	4	5	5	5
y_i	45	40	60	62	75	81	115	150	145	148

Table 1 . Scores vs. the number of training days

Our problem then is to use the proposed algorithm to find an optimal model fitting the data (minimizing the objective function $q(f)$).

B. Parameters

Our program is controlled by 6 parameters. In order to obtained effective parameter settings, a large number of computational tests with different combinations of parameters have been performed for the above test problem. Our best parameter settings obtained are listed below:

- The population size N is 100.
- The maximum number T of generations is 150.
- The probability of reproduction p_r is 0.10 .
- The probability of crossover p_c is 0.60 .
- The probability of mutation p_m is 0.30 .
- The maximum depth D of representation trees is 6.

Some of these parameters depend on the particular problem to be solved. For example, if the structure of the solution is thought to be more complex, a larger value of the maximum depth D might be appropriate. Our parameter settings are obtaianed by testing the following problem.

C. Results

For the data presented in Table 1, the following models (simplified by hand) are obtained from some runs:

$$
\begin{aligned}
f_1(x) &= 32.48 + 11.64\,x + (12.91 + 19.2\,x)\,e^{(1-8.22/x)} \\
f_2(x) &= 45.88 - 7.12/x + 4.16\,x^2 \\
f_3(x) &= 11.59 + 26.71\,x + 10.97\,cos(12.52 - x)
\end{aligned}
$$

To compare these models with the model $f(x) = 31.71 * (1.363)^x$ (obtained in [11]), we calculate their values of objective function $q(f) = \sum_{i=1}^{10} [y_i - f(x_i)]^2$, and their maximum absolute error $max_{ae} = \max_{1 \le i \le 10} |y_i - f(x_i)|$, minimum absolute error $min_{ae} = \min_{1 \le i \le 10} |y_i - f(x_i)|$, maximum relative error $max_{re} = \max_{1 \le i \le 10} \left| \frac{y_i - f(x_i)}{y_i} \right|$, minimum relative error $min_{re} = \min_{1 \le i \le 10} \left| \frac{y_i - f(x_i)}{y_i} \right|$, which are all listed in Table 2.

model	q	max_{ae}	min_{ae}	max_{re}	min_{re}
$f(x)$	103.13	5.56	0.71	0.081	0.0056
$f_1(x)$	96.60	5.77	0.23	0.094	0.0008
$f_2(x)$	91.83	5.93	0.07	0.079	0.0008
$f_3(x)$	80.06	5.73	0.06	0.094	0.0005

Table 2 . Comparison of models

From Table 2, the models $f_1(x), f_2(x)$ and $f_3(x)$ are all better than the model $f(x)$ in the meaning of least squares. Especially the model $f_2(x)$ is a prepolynomial model (those that can become a polynomial model through some transformations), whose coefficients can directly be optimized by the least squares method.

5. Conclusions

This paper presents an evolutionary computational approach to model-building problems, which is based on the ideas from both the evolution programs and the genetic programmings. The experimental results indicate that it is surprisingly effective in searching the optimal model for model-building problems through a quite "natural" representation procedure and some specific genetic operators. Because the genetic methods solve a problem by means of an evolving population of potential solutions, several near-optimal solutions can be obtained in a single run. It is essentially helpful for some practical problems such as regression analysis.

In comparison to traditional methods, a major advantage of this genetic method is that it can be unsupervised, i.e., one does not need to know the desired model forms and parameters. Both the model form and its parameters can be obtained in one run.

By the way, we test the algorithm also on some sets of data from given functions. If the given function is in F, then it can be successfully searched after some generations.

ACKNOWLEDGMENT The authors would like to thank the referees for their valuable comments and suggestions on this paper.

References

[1] D. E. Goldberg, Genetic Algorithms in Search, Optimization and Machine Learning, Addison Wesley, 1989.

[2] D. E. Goldberg and K. Deb, A Comparative Analysis of Selection Schemes Used in Genetic Algorithms, in G. Rawlins (Ed.), Foundations of Genetic Algorithms, 69–93,1991.

[3] J.H.Holland, Adaptation in Natural and Artificial Systems, The University of Michigan Press, 1975.

[4] J. R. Koza, Genetic Programming: On the Programming of Computers by Means of Natural Selection, MIT Press, Cambridge, 1992.

[5] H. Linhart and W. Zucchini, Model Selection, John Wiley & Sons, New York, 1986.

[6] V. Maniezzo, Genetic Evolution of the Topology and Weight Distribution of Neural Networks, IEEE Trans. on Neural Networks, vol.5(1), 39–53, 1994.

[7] Z. Michalewicz, Genetic Algorithms + Data Structures = Evolution Programs, Springer-Verlag, Berlin, 1992.

[8] T. Poggio and F. Girosi, Networks for Approximation and Learning, Proceeding of the IEEE, vol.78(9), 1481–1497, 1990.

[9] D. Whitley, The GENITOR Algorithm and Selection Pressure: Why Rank-based Allocation of Reproductive Trials is Best. in J. D. Schaffer (Ed.), *Proc.of the Third Int. Conf. on Genetic Algorithms*, 116–121, 1989.

[10] X. Yao, A Review of Evolutionary Artificial Neural Networks, International Joural of Intelligent Systems, vol.8(4), 539–567, 1993.

[11] X. D. Zhang, Applied Regression Analysis, Zhejiang University Press, 1991. (in Chinese)

Training Neural Networks With Influence Diagrams

Alexei Manso Corrêa Machado   Mário Fernando Montenegro Campos

DCC – ICEx – Universidade Federal de Minas Gerais
Caixa Postal 702 – Belo Horizonte – MG – CEP 30.161-970 – Brazil
e-mail: alexei@dcc.ufmg.br   e-mail: mario@dcc.ufmg.br

Abstract. This paper discusses the application of Influence Diagrams on training Neural Networks. The basic concepts of these two methodologies are presented as a brief review. The conventional back-propagation training procedure is compared to other alternatives, by means of an example on visual pattern recognition.

1   Introduction

The application of neural networks to pattern recognition problems has increased considerably, during the last 10 years. These mathematical models provide effective methods for complex problem solving that require robust, fault-tolerant solutions, that should be able to learn and generalize [1, 3]. Neural networks learn by examples and therefore need little previous knowledge. On the other hand, the set of examples used to train a network may be difficult to obtain. This set should be representative in such a manner that it favor generalization, without leading to the memorization of input/output pairs.

At the same time Neural Computation Theory comes back as an important research field, the use of Influence Diagrams (Bayesian Belief Networks) is broadly considered to decision analysis problems, motivated by the works of J. Pearl [6, 7]. Based on cause-effect relationships among the elements of the problem, this methodology allows the encoding of knowledge and uncertainty, constituting a powerful tool in problem solving such as medical diagnosis and investment risk analysis [2]. Pattern recognition can be viewed as a decision problem and therefore can be handled by influence diagrams. In [5] we have shown an example of using influence diagrams in Computer Vision. When handling populations with different probability distribution functions, its effectiveness is superior to Bayes Classifier [11] and comparable to neural network results.

How could we combine the advantages of these two methodologies to obtain a model that would offer both knowledge and learning? This paper describes the use of influence diagrams at the training phase of back-propagation neural networks. In section 2 and 3, a brief review of neural network and influence diagram theories is presented. Sections 4 and 5 are devoted to the analysis of an example and section 6 summarizes the main conclusions of this work.

2 Back-Propagation Neural Networks

The development of the back-propagation algorithm put an end to the limitations of simple perceptrons [8]. The introduction of intermediate "hidden" layers between the input and output layers has proved to be able to represent any boolean function. An example of a two-layer network is shown in Fig. 1. The algorithm uses a set of input/output patterns (x, y) to adjust the weights between units by means of gradient descent. Each pattern is applied to the input layer and the signal propagates to the hidden H and output O layers.

$$H_j = g\left(\sum_{k=0}^{N} w1_{jk}\, x_k \right), \quad \forall\, j = 1 \ldots Q \tag{1}$$

$$O_i = g\left(\sum_{j=0}^{Q} w2_{ij}\, H_j \right), \quad \forall\, i = 1 \ldots M \tag{2}$$

where N, Q and M are the number of units at the input, hidden and output layers respectively, $w1_{jk}$ is the connection weight between two units x_k and H_j, and $w2_{ij}$ is the connection weight between two units O_i and H_j. The **sigmoid** function can be used as the activation function g, as it must be differentiable.

$$g(t) = \frac{1}{1 + e^{-t}} \tag{3}$$

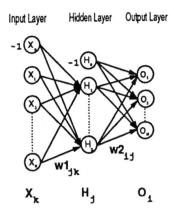

Fig. 1. An example of a two-layer neural network.

The desired pattern output y is compared to the output layer values, as a way of computing $\delta 2$. The same must be done to the hidden layer, so the errors are propagated backwards, producing $\delta 1$.

$$\delta 2_i = O_i(1 - O_i)(y_i - O_i), \quad \forall \, i = 1 \ldots M \tag{4}$$

$$\delta 1_j = H_j(1 - H_j) \sum_{i=1}^{M} \delta 2_i \, w2_{ij}, \quad \forall \, j = 1 \ldots Q \tag{5}$$

The connection weights can now be updated. The **learning rate term** η controls the speed of learning.

$$w2'_{ij} = w2_{ij} + \eta \, \delta 2_i \, H_j, \quad \forall \, i = 1 \ldots M, \quad \forall \, j = 0 \ldots Q \tag{6}$$

$$w1'_{jk} = w1_{jk} + \eta \, \delta 1_j \, x_k, \quad \forall \, k = 0 \ldots N, \quad \forall \, j = 1 \ldots Q \tag{7}$$

This procedure is repeated for each pattern μ in the training set, until a training epoch ϵ is completed. Many epochs may be necessary to adjust the weights. This can be verified by an error measure or cost function that goes to zero as the network produces "satisfactory" outputs.

$$E(\epsilon) = \frac{1}{2} \sum_{\mu,i} [y_i^\mu - O_i^\mu]^2 \tag{8}$$

It is also possible to accelerate the learning process by introducing a **momentum term** [3]. As shown, the training phase depends on a proper pattern set that is iteratively compared to the network output, as a way of adjusting the weights of the connections. The next section presents an alternative training method, that may substitute the training set in this process.

3 Influence Diagrams

Influence Diagrams (IDs) are directed acyclic graphs in which it is possible to represent causal influences between propositional variables [6, 7]. If there is a directed arc that goes from a node A to a node B, it can be said that variable A influences the values assumed by variable B, i.e. the probability of B assuming value b_j is **conditioned** by each value a_i of A. When more than one variable influences a node, the conditional probabilities will be a function of the state of its parents. Fig. 2 shows an example of a simple diagram. Node X is influenced by its parents A and B and influences its children Y and Z. The **fixed** conditional probabilities associated to X can be expressed as

$$P(x_i/a_j b_k) \quad \text{is the probability of } X = x_i, \text{ given that } A = a_j \text{ and } B = b_k. \quad (9)$$

The effectiveness of IDs is related to the correct specification of conditional probabilities. This can be accomplished by taking the histogram frequency of each feature as a measure of its occurance probability. The main disadvantage of this procedure is that it limits the space of each class to its pattern sample set, preventing generalization. Another way of determining probabilities is by using probability distribution models. Unlike in the case of the Bayes Classifier, distinct functions may be used for each population. This allows generalization and learning, but requires analytical functions which may not be possible to determine. Finally, the conditional probabilities can be posed by experts, as a way of encoding knowledge, restrictions and previous experience. When dealing with many features, however, this task may be too subjective and exhausting. In fact, these methods can be easily combined to get the best of each approach.

The second probabilistic measure in an ID is the **degree of belief** BEL. It reflects the overall conviction that a variable B assumes a value b_j, according to all incoming information, or **evidence e**, so far observed. As new evidence is introduced in the network, the belief of each proposition, that initially represents its *a priori* probability, must be updated. The degree of belief is **dynamic** in its nature and can be expressed as

$$BEL(x_i) = P(x_i/e) \quad \text{is the conviction that } X = x_i, \text{ given evidence e}. \quad (10)$$

IDs can be divided into two main topologies: **singly-connected**, if the graph presents only one path from a node to any other, or **multiply-connected**, if more than one path is found between two nodes. This division is essential to complexity analysis. While singly-connected diagrams present linear complexity behavior, multiply-connected ones may be exponentially hard [4, 9]. Fortunately, we find that many pattern recognition problems result in simple diagrams, where the choice of independent features leads to tree-like structures.

An ID is only a set of conditional probabilities, until some evidence is introduced in the network. This can be accomplished by **instantiating** a node, i.e. taking one of its values with maximum certainty. When this occurs, the overall network evidence is modified and the new information must be propagated to other nodes, in order to update the belief measures. The propagation is a chain process where special **messages** are sent from one node to its neighbors.

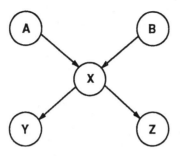

Fig. 2. A simple singly-connected influence diagram.

Any node may have m parents and n children, where $m \geq 0$, $n \geq 0$ and $m + n > 0$. We will call X a generic node, $P_{1...m}$ its m parents and $C_{1...n}$ its n children. Fig. 3 shows a fragment of an singly-connected ID. The **causal support** messages π are sent from one node to its children, while the **diagnostic support** messages λ goes from one node to its parents. The complete description of the propagation process in singly-connected networks can be found in [6, 7]. The general formulas for updating BEL, π and λ are given below, where α is a normalizing constant.

$$BEL(x) = \alpha[\prod_{j=1}^{n} \lambda_{C_j}(x)] \sum_{p_1 \cdots p_m} [\mathbf{P}(x/p_1 \ldots p_m) \prod_{i=1}^{m} \pi_X(p_i)] \qquad (11)$$

$$\pi_{C_j}(x) = \alpha[\prod_{k=1, k \neq j}^{n} \lambda_{C_k}(x)] \sum_{p_1 \cdots p_m} [\mathbf{P}(x/p_1 \ldots p_m) \prod_{i=1}^{m} \pi_X(p_i)] \qquad (12)$$

$$\lambda_X(p_i) = \alpha \sum_{\{p_1 \cdots p_m\} - \{p_i\}} [[\prod_{k=1, k \neq i}^{m} \pi_X(p_k)] \sum_{x} [\mathbf{P}(x/p_1 \ldots p_m) \prod_{j=1}^{n} \lambda_{C_j}(x)]]$$
$$(13)$$

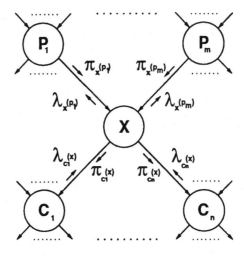

Fig. 3. A singly-connected influence diagram, with message parameters.

4 An Example

As an illustration of the use of IDs in pattern recognition applications, consider a simple instance of a more general problem of classifying plants by examining the shapes of their leaves. A sample of 144 leaves from 3 different plants — fern, hibiscus and jasmine — was captured as 100 ppi binary images, from which 2 shape features were extracted: the area and the compactness[1]. This two-dimensional feature vector has proved to be enough to achieve a reasonable separability of classes, besides allowing visual representation. Fig. 4 shows an example of a leaf of each class and the sample placement on the feature space.

An ID for this problem is composed of 3 nodes — Plant (P), Area (A) and Compactness (C) — and 2 arcs that links P to A and C (Fig. 5a). Because of conceptual independence between area and compactness, no arc links A and C, leading to a singly-connected diagram. In order to decrease the number of discrete values assumed by each feature node, they were grouped into 24 intervals. Node P can assume one of the 3 possible plant class values. The conditional probabilities $P(a_j/p_i)$ and $P(c_k/p_i)$ were easily determined by the normal distribution model, while the *a priori* probabilities $P(p_i)$ are proportional to the contribution of each class in the sample.

---

[1] *Compactness* indicates how circular an object is by relating perimeter P to area A, as $C = P^2/4\pi A$.

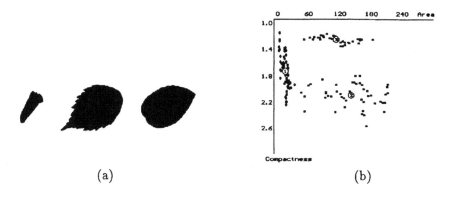

(a) (b)

Fig. 4. An example from each of the plant classes — 1-fern, 2-hibiscus and 3-jasmine (a). The sample placement on the feature space (b).

The decision function, in an ID, which settles that a pattern **x** Belongs to class ω_i, can be defined as

$$d_i(\mathbf{x}) = BEL(\omega_i) = P(\omega_i/e), \quad where \ \mathbf{x} \in \omega_i \ if \ d_i(\mathbf{x}) > d_j(\mathbf{x}) \ \forall j \neq i. \quad (14)$$

It should be pointed out that the degree of belief BEL reflects the conviction that a variable W assumes the value w_i, given the overall evidence e. In our example, the evidence is the feature values a_j, c_k extracted from pattern **x**.

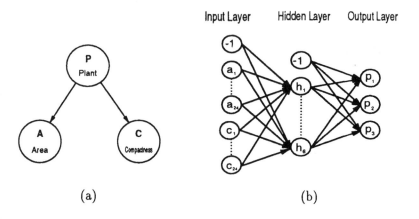

(a) (b)

Fig. 5. The Influence Diagram (a) and the Neural Network (b) from the leaves' example.

The corresponding neural network is shown in Fig. 5b. It has an input layer of 49 binary units (24 for each feature and one for threshold), a hidden layer of 7 units and an output layer of 3 units, one for each class value. The choice of binary units emphasizes generalization feature, although the input could have been encoded by 2 real-valued units.

In order to analyze the application of each method to the problem, five instances T_i were implemented, starting from the same initial random-weighted network, with learning rate of 0.4 and momentum of 0.9:

T1. Each pair of values (a, c) was applied to the influence diagram and the decision function (14) was used to determine its probable class. As seen in section 3, the values (a, c) are considered **evidences** and therefore must be propagated to update the BEL of node P. From (11), (12) and (13), the updated function BEL can be defined as

$$BEL(p) = \alpha \, \lambda_A(p) \, \lambda_C(p) \, \mathbf{P}(p)$$
$$= \alpha \, \mathbf{P}(a/p) \, \mathbf{P}(c/p) \, \mathbf{P}(p) \tag{15}$$

It should be noticed that the same function could be obtained by the direct application of Bayes' rule [10]. This is possible because we have a very simple diagram, which reduces the complexity of the propagation process. More complex problems may require including other types of information such as non-optical sensor measurements, leading to extended diagrams. In this case, the complete propagation process may have to be accomplished, generally in a recursive way.

T2. The 144 pattern sample was used to train the neural network with the conventional back-propagation algorithm. The deltas were determined by comparing the output with each pattern class (equations (4) and (5)). As the sigmoid function (3) never reaches 1 or 0, these values were substituted by 0.9 and 0.1, respectively, to denote pertinence and non-pertinence to a given class.

T3. Each sample pattern was applied to the ID and neural network. The deltas were determined by comparing the output from the network O to the BEL value of each class p, in the ID. Equation (4) becomes

$$\delta 2_i = O_i(1 - O_i)(BEL(p_i) - O_i), \quad \forall \, i = 1 \ldots M \tag{16}$$

This behavior can be explained by the fact that probabilities do not change abruptly when passing from one class region to another, at the feature space.

The ID's output provides more realistic measures, in this aspect, than by taking only binary values.

T4. Another approach to the problem is to substitute the sample by a set of random patterns as the training set. Thus, regions that are not exemplified at the sample can now be considered in the training phase.

T5. The training set can also be composed of patterns intentionally chosen to cover all the feature domain. This assures equal representativeness in the process.

5 Results

5.1 Separability and Generalization

The separability maps, resulting from the 5 situations described in section 4, are shown in Fig. 6. T1 shows how knowledge is encoded by the ID. The 3 classes are separated into 3 distinct areas and no pattern is misclassified. In T2, the conventional application of the back-propagation algorithm to the sample set creates "islands" where the absence of examples prevent the correct adjustment of connection weights. Feeding the network with the ID outputs, however, does not improve the generalization capacity of the network, leading also to discontinuous regions (Fig. 6c).

Better results can be obtained when the sample set is substituted by an randomly chosen pattern set with identical cardinality (Fig. 6d). This procedure reduces the number of disjoint areas, approaching the result obtained in T1. Special care must be taken when applying T4 in the training phase. Fig. 7 shows 3 ways of choosing patterns in a two-dimensional feature space. Example (a) seems to be well-distributed, but leads to poor separation of classes (Fig. 6e). Examples (b) and (c) are better, because all intervals are equally represented at the set. Fig. 6f shows the separability map of T5b, which seems to produce adequate results on separability and generalization.

5.2 Speed of Convergence

Fig. 8 shows the evolution of the cost function E, in the 20 first training epochs. No substantial difference is found with the application of T2 and T3, neither at T4 and T5. However, it can be pointed out that convergence is faster when a sample set is used, because patterns are concentrated in some regions, reinforcing weight updating. For well-distributed examples, more epochs are necessary to adjust the connection weights. This is however worth-while, since better separability results are obtained.

6 Conclusion

An alternative method for training neural networks was presented. The search for hybrid solutions can be prolific, if one is able to combine the main advantages of different approaches to produce better results. Neural networks are effective in learning and generalization. Influence diagrams can represent knowledge and human experience. The application of influence diagrams at the training phase of neural networks may substitute the requirement of training sets and transfer knowledge to them. The main aspects of this methodology were discussed in an example of pattern recognition. Although convergence speed may decrease in some cases, networks trained by influence diagrams produce superior results on generalizing concepts.

References

1. Davalo, E. & Naïm, P. *Des Réseaux de Neurones*. Eurolles, Paris, 1990.
2. Henrion, Max, Breese, J. S. & Horvitz, E. *Decision Analysis and Expert Systems*. AI Magazine, Winter 91.
3. Hertz, Jonh A. , Krogh, A. & Palmer, R. *Introduction to the Theory of Neural Computation*. Addison-Wesley, Redwood City, 1991.
4. Lauritzen, S. L. & Spiegelhalter, D. J. *Local Computations with Probabilities on Graphical Structures and Their Application to Expert Systems*. Journal Royal Statistical Society, 50, 1988.
5. Machado, A. C. & Campos, M. M. *On The Application of Influence Diagrams to Pattern Recognition Problems*. Proceedings of the Workshop on Cybernetic Vision, São Carlos, 1994.
6. Pearl, J. *Fusion, Propagation and Structuring in Belief Networks*. Artificial Intelligence, 29, 1986.
7. Pearl, J. *Distributed Revision of Composite Beliefs*. Artificial Intelligence, 33, 1987.
8. Rumelhart, D., Hinton, G. & Williams R. *Parallel Distributed Processing: Explorations in the Microstructures of Cognition*. MIT Press, Cambridge, 1986.
9. Suermondt, H. J. & Cooper, G. F. *Updating Probabilities in Multiply-Connected Belief Networks*. Proceedings of the 4th Workshop on Uncertainty in Artificial Intelligence, Minneapolis, 1988.
10. Thomas, J. B. *Introduction to Probability*. Springer-Verlag, New York, 1986.
11. Tou, J. T. & Gonzalez, R. *Pattern Recognition Principles*. Reading, 1974.

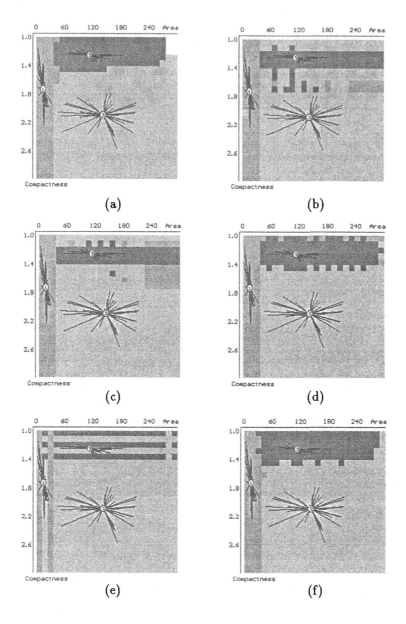

Fig. 6. The separability maps from T1 (a), T2 (b), T3 (c), T4 (d), T5a (e) and T5b (f).

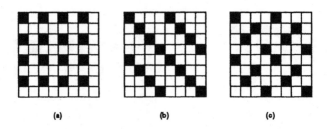

Fig. 7. Three ways of choosing patterns in a two-dimensional feature space, with density 0.25.

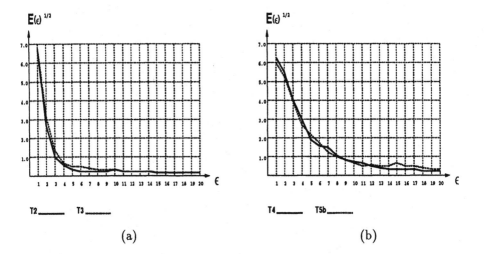

(a)   (b)

Fig. 8. The square root of the cost function **E** at each training epoch ϵ, for T2, T3 (a) and T4, T5b (b).

A Behavioural Theory of Intelligent Machines as a Framework for the Analysis of Adaptation

Craig A. Lindley

CSIRO Division of Information Technology
Locked Bag 17, North Ryde NSW Australia 2113

Abstract. This paper presents a novel Behavioural Theory of Intelligent Machines (BTIM) as a method of analysis within the behavioural paradigm of artificial intelligence, with a particular emphasis upon adaptive and self-organising processes. The BTIM provides normative metrics for understanding structural characteristics of systems, and supports comparisons across applications that might otherwise appear to be incommensurable. To demonstrate the use of the BTIM, a comparative analysis of two evolutionary systems is presented.

1   Introduction

This paper presents a Behavioural Theory of Intelligent Machines (BTIM) as a behaviourist's reply to what will here be referred to as the "IPDI Theory of Intelligent Machines" (ITIM) proposed by Valavanis and Saridis in [10]. Both the ITIM and the behavioural approach to artificial intelligence have been developed in the context of robotic systems. However, robot architectures manifest principles of agent design that can also be applied in general-purpose information systems when computational processes are modelled in terms of an agent or multi-agent metaphor. The BTIM incorporates many of the methodological tools of ITIM, but reinterprets those tools from the perspective of the behavioural paradigm of artificial intelligence, with a particular emphasis upon adaptive and self-organising processes. In particular, the BTIM provides a number of novel metric definitions, including an overall normative metric for the *merit* of autonomous agent designs that captures design characteristics that behavioural approaches implicitly seek to maximise. The BTIM provides normative metrics for the analysis and comparative evaluation of behavioural and adaptive systems.

2   General Systems Theory

Both the ITIM and the BTIM use concepts from Conant's general systems theory (see [4]) in which Information Theory is used to analyse general systems. Information Theory (as introduced by Shannon) is the study of the communication of messages between points. N-dimensional Information Theory has major advantages over other techniques of statistical analysis (see [10]). In particular, it supports quantifications of the degree to which different parts of a system are in communication with one another, it can measure rates of constraint between dynamic variables while taking account of past history, and the additivity of information measures allows the decomposition of information constraints among system variables. Conant in [4] has used entropy and transmission rates to define measures for dynamic systems that can be used for tradeoff analyses and to place lower bounds on the components needed to accomplish a task. Because an entropy representation of a random process is related to mean square error estimation theory, the entropy function can be applied to the class of systems that can be analysed using mean square error estimation techniques, and any system can be optimised by minimisation of its total entropy (see [10]). The application of general systems theory to the analysis of adaptive algorithms will be illustrated in this section by relating the theory to principles of evolutionary computation.

The following notational conventions are from Conant in [4]. S and $X_1, \ldots, X_n$ are used interchangeably to denote either a *set* or a *vector* of *system variables*. E is a *vector input* to a system S from its *environment*. E is considered to be all relevant variables not in S. $S_0 = X_1, \ldots, X_k$ are those variables X_j in S that can be directly observed from the environment, constituting *output variables*. $S_{int} = X_m, \ldots, X_n$ are the remaining *internal variables* X_j in S.

The basic measure associated with a discrete random variable X is the entropy H(X) of the channel represented by X as an information source. H(X) is a function of the probability distribution of X, and is defined as:

$$H(X) \quad = \quad - \sum_x p(x) \log p(x) \qquad (1)$$

where the summation is over all values x taken by X, and p(x) is the probability distribution function of X (see [4]). The entropy is the average uncertainty of particular values x taken by X, where uncertainty is measured by $- \log p(x)$. H(X) is the average amount of information obtained, given the value x taken by X on an arbitrary occasion, if the distribution p(x) is known *a priori*. The base of the log determines the units of the entropy measure ("Nats" for natural logs with base e). If X has a finite number n of values, H(X) falls within the interval $[0, \log n]$. Low values of H(X) within this range indicate a concentrated probability distribution function (pdf), and high values indicate a diffuse pdf. H(X) = 0 if and only if some value of x has probability 1, and H(X) = log n if and only if all $x \in X$ are equally probable.

The *entropy of a system of variables*, H(S), is defined by:

$$H(S) \quad = \quad - \sum_s p(s) \log p(s) \qquad (2)$$

where the summation is over all possible values s of the vector $S = X_1, \ldots, X_n$, and p is the distribution on the n-tuples describing S.

Conditional entropies, such as $H_{X1,X2}(X_3)$ indicating the average uncertainty remaining about X_3 when X_1, X_2 are known, can be defined either through the conditional distribution or through entropies:

$$H_{X1,X2}(X_3) \quad = \quad - \sum_{x1,x2,x3} p(x_1,x_2,x_3) \log p(x_3|x_1 x_2) \qquad (3)$$

$$= \quad H(X_1,X_2,X_3) - H(X_1,X_2) \qquad (4)$$

Entropies are *additive*, so that $H(X_1,X_2,X_3) = H(X_1,X_2) + H_{X1,X2}(X_3)$, which can be interpreted as partitioning the uncertainty of $H(X_1,X_2,X_3)$ into the uncertainty of X_1, the uncertainty of X_2 given knowledge of X_1, and the uncertainty of X_3 given knowledge of X_1 and X_2.

The *transmission* between two variables is a measure of the amount by which the variables are related (ie. not statistically independent). That is, it is a measure of the amount by which information about one variable reduces uncertainty about the other. The transmission, denoted $T(X_1:X_2)$, is defined through probabilities, or by:

$$T(X_1:X_2) = \qquad H(X_1) + H(X_2) - H(X_1,X_2) \qquad\qquad (5)$$

$T(X_1:X_2)$ falls in the interval $[0, \min\{H(X_1),H(X_2)\}]$, being zero if X_1 and X_2 are statistically independent, and maximum if one variable completely determines the other. Generalised transmission, $T(X_1:X_2,: \ldots :X_n)$, measures the global relatedness or constraint in a system of n variables, and constitutes an upper bound on the transmission between disjoint subsets of the n variables. Transmission between subsystems S^i, viewed as vector-valued variables, is similarly defined. Transmissions may be conditioned on a particular variable or vector. All transmissions and conditional transmissions are additive and non-negative, allowing the total relatedness or constraint between groups of variables to be partitioned analogously to the partitioning of entropy or uncertainty.

Conant in [4] notes that entropy and transmission rates do not account for historical context, so historical context cannot be used to reduce uncertainty. This deficiency is shared by all statistical techniques that use data as a collection of dissociated descriptions of the system. The problem can be overcome by defining the *entropy rate*, $\mathbb{H}(X)$, (with units, for example, in Nats per step), which is the entropy of X conditional upon all of its prior values, representing the information carried per observation in a long sequence. The total uncertainty of a long sequence $<X(t), X(t+1), \ldots , X(t+m)>$ is then (approximately) the entropy rate times the sequence length. The entropy rate is useful for measuring nondeterministic or unpredicatable behaviour of a variable or system.

The *transmission rate* of a system, $\mathbb{T}(S)$, measures the constraint per step holding between dynamic variables, defined by the entropy rate of each variable minus the joint entropy of the variables. The transmission rate is useful for measuring the relatedness of variables or subsystems. As Conant observes in [4], $\mathbb{T}(\text{environment} : \text{behaviour})$ measures information flow between the environment and behaviour that is coordinated with that information, and is therefore an important survival parameter of an agent and a limit on its ability to cope with dynamic circumstances. From a evolutionary computational viewpoint (not explored by Conant), this suggests that the convergence of an evolutionary process towards an optimal solution as defined by a fitness function is revealed by an increase through successive generations in the population average value of $\mathbb{T}(\text{environment} : \text{behaviour})$. Particular evolutionary methods can also be characterised in terms of an average or typical value of $\mathbb{T}(\text{generation } i : \text{generation } i+1)$, understood as the reduction in H(generation i+1) based upon performance information represented by H(generation i). This particular transmission rate corresponds to the rate of convergence of a particular algorithm in relation to a given problem.

Total rate of information flow , F, is defined by the sum of the entropy rates for all system variables. *Throughput rate* , F_t, measures the information flow rate of S, and is defined as the transmission from the environment variables, E, to the output vector of the system, S_o. Survival as a goal usually dictates some minimum value of F_t. *Blockage rate* , F_b, measures information carried by the internal state variable set, S_{int}, about the environment E that is not measured in the output. F_b corresponds to the amount of filtering out of (irrelevant) information. The blockage rate for evolutionary processes measures the average loss of unfit individuals in a population through successive generations, where a generation i is regarded as (a subset of) the input to a generation i+1; the information carried by unfit individuals that is filtered out is information about the space of possible solutions that does not have relatively high adaptive merit. *Coordination rate*, $F_c = \mathbb{T}(X_1:X_2: \ldots :X_n)$, can be regarded as the rate of internal communication within S that allows S to act as a coordinated whole, rather than a sum of independent parts. It is defined as the transmission between the complete set of system variables. Conant suggests in [4] that the minimum F_c needed to successfully cope with a problem can be (loosely) interpreted as corresponding to the complexity of the problem in terms of the amount of global cooperation required between parts of S for its successful solution. Hence, F_c can be used as a comparative measure of the complexity of alternative solution techniques for a given problem. Within evolutionary processes, coordination rates can be used to model crossover rates within each generation. *Noise rate*, F_n, is the amount of uncertainty per step about S, given complete knowledge of its input. F_n can be used to measure stochastic elements of an evolutionary process, such as the random selection of individuals for genetic operations, or the selection of crossover points within individuals. A *deterministic* system is one for which all rates conditional upon E are zero, so if complete information of the system inputs is available, then there is no residual uncertainty about any of the system variables, and hence they carry no information either individually or about each other. F_n is zero in the case of a deterministic system.

The *Partition Law of Information Rates* (PLIR) is expressed by:

$$F \quad = \quad F_t + F_b + F_c + F_n \tag{6}$$

F represents the total (nondeterministic) activity in S if intervariable relationships are ignored. Each X_j is assumed to be a (possibly noisy) function of variables in E and S. F can be viewed as an entity that computes that activity level, and therefore (loosely) represents the amount of computation carried out within S. F bounds F_t, F_b, F_c, and F_n individually and cumulatively, and therefore F represents a global upper bound for all information rates. PLIR does not imply a partition of system variables corresponding with the particular subrates of F; those subrates represent global measures on the system that cannot be localised. For a deterministic system, $F_n = 0$, and the PLIR becomes $F = F_t + F_b + F_c$, which may be called the *Deterministic Partition Law of Information Rates* (DPLIR). The output of the system in this case is equivalent to its throughput, so F_t can be considered the output rate.

The PLIR clarifies tradeoffs between throughput, blockage, and coordination for a fixed or bounded total rate. Demands of F_t, F_b and F_c are essentially independent and additive in the requirements they place on the information-processing or computing resources (F) of the system. The noise rate, F_n, competes with all other rates when F is fixed, and so is typically minimised in order to maximise any other rate. On this basis, Conant concludes in [4] that efficient processors (and, presumably, processes) must be deterministic. Adaptive processes must explore the state space of a system, and the exploratory component can be modelled as noise. It follows that an adaptive process is not maximally efficient in Conant's sense. However, adaptive processes *can* in principle achieve an optimal ratio of exploratory activity to exploitative activity (see [6]). In general, F is a constant upper bound upon system activity. Consider an adaptive process that is known to achieve the optimal ratio of exploration to exploitation, where F_n is due entirely to exploration, the system is operating at maximum information capacity, and the environment dynamics are stable. In this case, as adaptation proceeds F_n will approach (but never quite reach) 0, while the sum $F_t + F_b + F_c$ will approach F.

3 The IPDI Theory of Intelligent Machines

The IPDI Theory of Intelligent Machines (ITIM) has been proposed by Valavanis and Saridis in [10] based upon an analytic formulation of the principle of Increasing Precision with Decreasing Intelligence (IPDI). Valavanis and Saridis define *intelligent machines* as "hierarchical structures in the order of increasing intelligence and decreasing precision". Valavanis and Saridis use the IPDI to justify three primary hierarchical partitions of the architecture of intelligent machines, with further hierarchical decomposition within each primary layer. The resulting architecture is an example of what may be called a "representational" intelligent robot control architecture. Representational systems use symbolic models of the robot, its operating environment, and the range of tasks and actions that the robot "knows how to perform". System behaviours are produced by an inference engine that reasons about the models to produce action plans (Georgeff reviews planning within the representational paradigm in [5]). Plans and action representations are typically hierarchically ordered, with high level representations of goals and actions at the top of the hierarchy. Subsequent levels of abstraction decompose goals into increasingly specific subgoals, ending with primitive machine commands that can be executed directly by hardware. Robotic plans are monitored during execution. The robot may attempt to deal with plan failures by replanning from some suitable level of abstraction, with various techniques being employed to deal with constraints of real-time operation.

The formulation of the principle of IPDI by Valavanis and Saridis (see [10]) is an attempt to use the concept of entropy as a unifying principle for intelligent systems conceived of as entropy-reducing systems. IPDI is based upon a number of definitions that use the concept of entropy. *Machine Knowledge* (K) is defined informally by Valavanis and Saridis in [10] as "the structured information acquired and applied to remove ignorance or uncertainty about a specific task pertaining to the Intelligent Machine". This definition includes both *a priori* and *a posteriori* knowledge, and allows for knowledge to be represented explicitly (eg. as data or knowledge bases) or implicitly (eg. embedded within control or learning algorithms). Machine Knowledge indicates the total amount of information that has been accrued in the long term memory of the organisational level of a system. Valavanis and Saridis in [10] present a formal definition of knowledge that is far less general. The total knowledge of a system, having the form of an energy of the underlying events, is defined as:

$$K = 1/2 \sum_i \sum_j w_{ij} s_i s_j \qquad (7)$$

where $s \in \Omega_s$ represent the state of events at the nodes of a network defining the stages of a task to be executed, and w_{ij} are the state transition coefficients between states (equal to zero in the case of inactive transmission). The form of this definition fundamentally associates knowledge with representation, since it depends upon the representation of tasks in terms of explicit stages.

Rate of Machine Knowledge ($R = K/T$, where T is a fixed time interval) is defined as the flow of knowledge through an Intelligent Machine. *Machine Intelligence* (MI) is defined as "the process of analysing, organising, and converting data into knowledge. Machine Intelligence, is the set of actions or rules which operates on a Data Base (DB) of events or activities to produce flow of knowledge (R)". *Machine imprecision* is defined as the uncertainty of execution of the various tasks of the Intelligent Machine". *Machine precision* is defined as "the complement of machine imprecision and represents the complexity of the process".

The principle of IPDI is expressed probabilistically by the relation: Prob(MI,DB) = Prob(R), where MI is the machine intelligence, DB is the database associated with the task to be executed, and R is the knowledge rate. DB represents the complexity of the task, and is proportional to the precision of execution. The corresponding entropy equation is: H(MI|DB) + H(DB) = H(R). A constant rate of knowledge is assumed during the conception and execution of a task. Hence, increasing the entropy of DB requires a decrease of the entropy of MI for a particular database, which manifests the principle of IPDI.

Valavanis and Saridis use the principle of IPDI to justify a hierarchical decomposition of system functions involving three primary hierarchical levels. The *Organisation* level is conceived in terms of the structure and principles of a traditional knowledge base or expert system. This level uses forward and/or backward chaining to generate plan scenarios, using either probabilistic or deterministic representation and plan generation methods. The *Coordination* level consists of a number of coordinators and a dispatcher. Each coordinator has a fixed structure, and is associated with a particular hardware subsystem or process. Communication between coordinators is mediated by the dispatcher. The dispatcher configuration is dictated by the organiser. Feedback from the execution level allows the coordination level to modify execution scenarios in real time. The dispatcher also communicates feedback to the organiser. The *Execution* level consists of hardware controllers that drive mechanisms and sensor processing systems.

The system is viewed as transforming a qualitative user command received by the organisation level through several hierarchical decompositions to a sequence of mechanical actions. The organisation level satisfies a command by organising an appropriate sequence of abstract actions, general tasks, and rules from a set of primitives stored in a long term memory. The resulting plan is passed to the coordination level which takes real time information about the world into account in generating a sequence of subtasks that can execute the plan. Coordination level functions include the selection of one among alternative plan scripts that accomplish the same task in different ways, according to workspace constraints and timing relationships. Subtasks are then executed by the execution level issuing actuator commands to appropriate hardware devices. Performance measures for each level are expressed as entropies, and this is taken by Valavanis and Saridis to unify the functions of the Intelligent Machine. Valavanis and Saridis take the principle of IPDI to apply across an individual hierarchical level of an Intelligent Machine (in which case R is held to represent information throughput in an information theoretic sense), as well as through the levels of the hierarchy.

4 A Critique of The IPDI Theory

There are three grounds for questioning the universal validity of the IPDI Theory of Intelligent Machines. Firstly, the principle of IPDI depends crucially upon the constancy of the knowledge rate, R. By definition, a constant R means a constant flow of structured, uncertainty-removing information during the conception and execution of a task. An information-theoretical analysis of system performance does *not* require the constancy of any particular rate, although the total rate of information flow, F, is bounded for a given system. If hierarchy is ignored, R is equated by Valavanis and Saridis with F_t, which is not constant, although it does have the upper bound of F.

Secondly, if under some interpretation the IPDI entropy equation, H(MI|DB) + H(DB) = H(R), holds, this does not necessarily imply any hierarchical decomposition beyond that implicit in the distinction already drawn in the definitions of MI and DB. By composition of the definitions presented by Valavanis and Saridis in [10], Machine Intelligence (MI) is defined as the set of actions or rules which operates on a database of events or activities (DB) to produce the flow of structured information acquired and applied to remove uncertainty about a specific task. The entropy equation alone has no specific implications for the substructure of that set of actions or rules constituting MI, nor can any conclusions be drawn about that substructure from the natural language definition of Machine Intelligence.

Finally, the Theory of Intelligent Machines uses hierarchical task decomposition as the primary structuring principle, and task representation and deliberation within the organisation level as a basis for performing high level functions. Valavanis and Saridis acknowledge in [10] that Intelligent Machines share the lack of common sense knowledge and the inflexibility of expert and knowledge based systems, and acknowledge the inability of the architectural model presented to explicitly consider the possibility of unpredicatable events occurring during plan execution. Some limited methods are available for dealing with failures due to command misclassification, due to insufficient primitive tasks within a plan to generate basic task functions, due to incorrect use of basic task functions, and due to insufficient basic task functions. However, the problems of recognising when any of these errors has occurred, recognising which error has occurred, and then of producing appropriate corrective action within the real time constraints of system operation, are as great in this case as for any other representation-centred deliberative planning system, since these systems can only handle problems that fall within the scope of their modelling capacity.

5 General Systems Theory and Behavioural AI

Behavioural approaches to the development of robotic systems have directly addressed problems encountered by representational systems by producing "intelligent" behaviour without depending upon high levels of knowledge representation (see [9]). Instead, inspired by principles manifested in biological systems, agent architectures are structured to produce behaviours directly, using the simplest possible implementation methods. For example, the Subsumption Architecture (see [1]) is based upon parallel layers of behaviour-producing networks of augmented finite state machines. No world models are used, and the resulting systems demonstrate superior performance in real applications. However, more rigorous behavioural methodologies are needed for the analysis, design, and comparative evaluation of intelligent behavioural systems (see [2]). The designer of an autonomous agent or system is faced with many questions regarding the robustness, complexity, and adaptivity of the system. Frequent claims are made regarding the superior flexibility and robustness of behavioural systems in comparison with traditional systems, but these claims are rarely quantified; within the behavioural paradigm there are no common metrics used for the evaluation of the difficulty of a particular problem or task domain, or of the performance of particular agent designs. It is proposed here that Information Theory and Conant's general systems theory can provide a number of useful metrics for addressing these questions if the questions can be formulated probabilistically. This section considers some relationships between general systems theory and behavioural approaches to artificial intelligence, as a prelude to a more formal definition of a Behavioural Theory of Intelligent Machines (BTIM).

Conant argues in [4] that a hierarchical structure is optimal for the maximum operation of system components at full capacity, since a hierarchical architecture allows (but does not necessarily guarantee) the operation of every component in S at its full capacity, minimising the mere point-to-point transfer of information and maximising the prompt blockage of irrelevant input information. The hierarchical decomposition rule for $\mathbb{H}$ is:

$$\mathbb{H}(S) \quad = \quad \sum_{i=1}^{N} \mathbb{H}(S^i) - \mathbb{T}(S^1 : S^2, : \ldots : S^N) \tag{8}$$

This amounts to the rule that the entropy rate of a system S is equal to the sum of the entropy rates of its subsystems, S^i, minus the transmission rate between those subsystems. Decomposition can be applied iteratively to each subsystem, resulting in a total hierarchical decomposition. If the transmission rate of information between system variables *within* subsystems is much larger than the transmission of information *between* subsystems, then S is decomposable into nearly independent subsystems that can be understood in near isolation. This important principle underlies hierarchical decomposition as a method for dealing with complexity by decomposing complex systems into nearly independent, less complex, and modular subsystems. The concepts of flow, throughput, blockage, and coordination rates can similarly be decomposed into corresponding metrics for subsystems minus the values accounted for by subsystem interrelationships.

However, the decomposition rules as stated do *not* necessarily entail a hierarchy in any sense other than a hierarchical system of abstraction for viewing the structure of a complex system. In particular, the decomposition rules do not in any way dictate a hierarchical decomposition of procedures or control sequences for achieving the aims of the system. Hence Conant's general systems theory is not intrinsically incompatible with approaches that do *not* depend upon a hierarchical structuring of control or functional flow. Incorporating Conant's metrics into a Behavioural Theory of Intelligent Machines (BTIM) involves a reinterpretation of the kind of measures defined by Valavanis and Saridis in [10] such that hierarchy is no longer the dominant organisational principle of an autonomous control system.

A characterisation of systems based upon Information Theory provides performance, efficiency, and complexity criteria for choosing between alternate system designs. In particular, among systems that can achieve the desired operational goals of a particular application, designs that minimise F can be chosen. Systems that minimise F will be those that minimise total system channel capacity, and (loosely) the number of variables or components in S. F_b depends more upon E than upon S. This suggests that F should be minimised by reducing F_c to the minimum consistent with satisfying the operational goals of the system. Hence, the subparts of S should operate as independently as possible, and the best substructure of S is the loosest structure compatible with satisfying the goals of the applications. In general, Conant concludes in [4] that solutions should be selected that:

1. produce the minimum allowable output (hence, do not produce any unnecessary output).

2. perform as little blockage as possible (hence, accept the minimum irrelevant input).

3. reduce internal coordination to the minimum consistent with the other requirements (hence, maximise the freedom of the components of the system).

4. as far as possible, match components to tasks so that each component is operated at maximum capacity.

The behavioural paradigm has independently converged towards system design principles embodying these conclusions of Conant's theory, driven by the practicalities of building real artificial agents operating in real environments.

6   A Behavioural Theory of Intelligent Machines

This section introduces novel formal definitions of a minimum number of simple BTIM concepts to demonstrate the application of entropy-based concepts in relation to two simple adaptation cases.

Maes in [9] has described the following characteristics of behavioural approaches that distinguish them from representation-centred approaches:

- systems have multiple, integrated, and typically low-level competences

- the system is "open" or "situated" in its environment. The system has many interactive interfaces with its complex, dynamic, and unpredictable environment

- the emphasis is upon autonomy

- systems are designed to produce behaviour, rather than to have knowledge in a representational sense. Particular functions may emerge from the activity of more primitive behaviours.

- there is a strong emphasis upon adaptation

Any particular system can be characterised by an arbitrary subset of these features, so a behavioural approach is to some extent a matter of degree. Lindley has argued in [8] that behavioural approaches are justified by philosophical positions that regard represented knowledge as a cultural and pragmatic artefact, rather than being the substance of intelligence; in effect, knowledge is reified practice, codified by representation in order to coordinate behaviour. From this viewpoint, *behaviour* is at the core of intelligence. The features listed above tend to follow naturally from this perspective. There are many different types of behavioural systems, and it necessary to formalise a definition based upon typical features, upon a philosophical perspective, or upon case studies of behavioural systems. One universal theme of behavioural approaches is the rejection of world modelling as the only path to intelligence, raising new questions about fundamental approaches to artificial intelligence. A generalising definition of the behavioural paradigm should manifest this inquiring attitude, and should not be tied to any particular techniques or metaphors for achieving intelligence (at the same time as not excluding any particular techniques *a priori*). A novel formal definition will be developed here that captures this technique-neutrality, while providing a quantitative measure of merit for autonomous systems. The development begins with several preliminary definitions.

Adaptation can be defined for a system variable X as the capacity of an agent (or subsystem) to reduce the entropy of X. That is, the adaptation of X is defined as:

$$A(X) \quad = \quad H(X_{initial}) - H(X_{final}) \tag{9}$$

A corresponding average *adaptation rate* can be defined as:

$$\Delta A(X) \quad = \quad A(X)/\Delta t \tag{10}$$

where Δt is the time interval (or number of time steps) over which the entropy change described by $A(X)$ occurs. $\Delta A(X)$ differs from Conant's definition of entropy rate in that $\Delta A(X)$ is the average rate of entropy reduction of X during the period Δt, whereas the entropy rate, $\mathbb{H}(X)$, is simply the average entropy of X over an indefinite time period.

Adaptive efficiency in relation to a system variable X is defined as the adaptation divided by the complexity, C, of the design:

$$E_A(X) \quad = \quad A(X)/C \tag{11}$$

Design complexity can be measured in numerous ways, and can incorporate both computational time complexity and computational space complexity. Here it is simply the number of atomic language components used to express a design (eg. the number of function and terminal names in a LISP S-expression).

Adaptation efficiency rate in relation to a system variable X is defined as the adaptation efficiency divided by the time taken for each unit of compexity to achieve a given level of adaptation:

$$\Delta E_A(X) \quad = \quad A(X)/C\Delta t \tag{12}$$

With these definitions, the *Behavioural Theory of Intelligent Machines (BTIM)* can now be defined as: a systematic research program that aims to maximise the merit of autonomous systems, where the *merit* of a system S is defined as:

$$\text{merit}(S) \quad = \quad \left(\Sigma_{x \in S} \Delta E_A(X)\right) / F_c \tag{13}$$

$$= \quad \Sigma_{x \in S} A(X)/F_c C\Delta t \tag{14}$$

$$= \quad \frac{\Sigma_{x \in S} [H(X_{initial}) - H(X_{final})]}{F_c C\Delta t} \tag{15}$$

with units of atoms^{-1}, where F_c is the coordination rate as defined by Conant in [4]. The merit metric can be used in this complete form, or systems can be analysed in terms of the component expressions and their proportionate effect upon merit.

The overall merit, the adaptation, or the adaptation rate of a system, are related to the *intelligence* of the system. However, the somewhat arbitrary association of emotive terms with formal definitions adds little light to the problem of understanding complex adaptive systems, so intelligence will not be formally defined as part of the BTIM. *Machine knowledge* is defined by Valavanis and Saridis in [10] as the "structured information acquired and applied to remove ignorance or uncertainty". By this definition, machine knowledge is distinguished from the basic concept of information by reference to an intrinsic structure. To account for this within the BTIM, machine knowledge is equated with the sum of transmission relationships between the variables of a system; that is, machine knowledge is what Conant refers to in [4] as *generalised transmission*. This definition has the advantage of neutrality with respect to the computational languages and models used to implement uncertainty-reducing relationships, and so is not committed to ideas of high-level representation and modelling.

Behavioural approaches to artificial intelligence can now be characterised as those approaches at least implicitly concerned with maximising the BTIM metric of merit. A typical behavioural agent architecture does not have a predominantly hierarchical structure, but is decomposed at the top level into a series of (hypothetically) minimally coupled subsystems (in the sense defined by Conant in [3]), such as the layered competencies of the subsumption architecture (see [1]). Hence, the system architecture is *not* structured according to representations embodying levels of abstraction manifesting reducing entropy. Rather, the architecture is structured at the highest level to reduce the coordination rate, F_c, and thereby increase the merit of the system as measured by the proposed quantitative metric. The elimination of centralised world modelling in a behavioural system is critical to the minimisation of F_c, since the world model itself represents a major nexus of coordination (typically accounting for most of

F_c in a representation-centred system). Minimal coupling between functional modules is a paradigmatic characteristic of behavioural automata (see [9]).

7 Case Studies

In this section a simple comparison of the performance two adaptive systems is presented to show how BTIM metrics can be used to compare adaptive automata having quite different forms and functions. Koza in [7] describes the use of genetic programming (GP) for learning simple control rules for behavioural agents that achieve collective emergent functionality, and for learning the control rules for a single behavioural robot organised according to the subsumption architecture. GP is based upon ideas from Darwinian evolution, incorporating computational models of fitness-proportionate reproduction and adaptation by genetic crossover. It is generally a robust method that Koza shows to be capable of handling a degree of inaccuracy in the fitness measure used, incomplete environmental information, catastrophic damage to the population, and unanticipated environmental changes.

7.1 Adaptation by Genetic Programming

GP manipulates general, hierarchical computer programs of dynamically varying size and shape. The search space is therefore the space of all possible computer programs composed of a set, F, of functions and a set, T, of terminals appropriate to the problem domain and satisfying requirements of closure and sufficiency. *Closure* requires that each function should be able to accept as arguments any value that may be returned by any function in the function set and any value or data type that may possibly be assumed by members of the terminal set. If closure does not hold, other mechanisms must be established to eliminate programs (possibly after creation) that violate typing constraints. *Sufficiency* requires that the set of functions and terminals be capable of expressing a solution to the problem.

The GP algorithm is a method for searching the space of possible programs to find the best (or fittest) program to solve a particular problem. Fitness is generally assessed by running a candidate program over a number of *fitness cases* representative of the problem space. Fitness is primarily stated in terms natural to the problem, and commonly has the form of an error measured as the difference between some target performance value and the actual performance value summed or averaged over the set of fitness cases. The fitness function can also take into account secondary or tertiary factors such as parsimony or efficiency of the program.

The basic GP algorithm is:

1. Generate an initial random population of compositions of functions and terminals of the problem (ie. computer programs).

2. Iteratively perform the following substeps until the termination criterion has been satisfied:

    a. Execute each program in the population and assign a fitness value to each according to how well it solves the problem.

    b. create a new population of computer programs by applying the two following primary operations. The operations are applied to computer programs in the population chosen with a probability proportionate to their fitness:

        i. Copy existing programs to the new population ("asexual reproduction").

        ii. Create new computer programs by "genetically" recombining randomly chosen points of two existing programs ("sexual reproduction").

3. The best computer program that appeared in any generation (ie. the best-so-far individual) is designated as the result of the algorithm.

In asexual reproduction, one parent program per operation is reproduced without modification. The parent is selected according to a probability equal to its fitness divided by the sum of the fitnesses of all members of the population (alternative selection methods, such as rank selection and tournament selection, are not considered here).

The sexual crossover operation begins with two parents and produces two offspring. Both parents are chosen using the same fitness-based selection method used for reproduction. Using a uniform probability distribution, a random point in each parent is independently selected as the crossover point for that parent. The subtrees in each hierarchical program structure beginning at each respective crossover point are then swapped between the parents. The closure property ensures that the two resulting offspring programs are legal expressions in the language used (LISP S-expressions in [7]). A number of secondary operations may be performed, including mutation, permutation, editing, encapsulation, and decimation. Empirical results presented by Koza in [7] suggest that these secondary operations have little effect upon performance, although decimation can correct a population skewed towards poorly performing individuals.

A run of genetic programming can be controlled by a number of parameters. The two primary parameters are the population size and the number of generations to create. Secondary parameters include the probabilities and/or frequencies of secondary operations, and size constraints upon individuals. A run can be terminated either when the specified maximum number of generations has been created, or when a program has been created that satisfies a specified success criterion.

7.2   Learning Emergent Behaviours by Genetic Programming

To demonstrate the use of BTIM metrics, this section considers the application of Genetic Programming to the task of learning *emergent central place food foraging* behaviour. The GP experiment is described by Koza in [7]; here the results of Koza's work are analysed by application of the BTIM. In the emergent central place food foraging problem, the goal is to genetically breed a common computer program that, when simultaneously executed by all of the individuals in a group of independent agents, causes the emergence the beneficial and interesting higher-level behaviour of transporting available food to a nest or colony. The agents in the example are simulated ants. Ants intercommunicate indirectly via their environment, leaving pheremones along a trail from food back to the nest so that other ants can follow a common trail to the food. Each ant leaves a trail of pheremones that disperse over time, so the trail is only maintained after initial discovery of the food for as long as the food lasts and ants continue to travel along it.

Koza's experiment involves a simulated world consisting of 144 food pellets piled eight deep in two 3 x 3 piles on a 32 x 32 toroidal grid. There are twenty ants. The state of each ant is defined by a grid position, a facing direction, and an indication of whether the ant is currently carrying food. Each ant begins at rest facing in a random direction at a random location. Each ant has a copy of the common computer program, and the program is executed for each ant at each simulation time step.

There are nine operators used to define the program, namely the terminal set (with arity 0): MOVE-RANDOM, MOVE-TO-NEST, PICK-UP, and DROP-PHEREMONE; and the function set: IF-FOOD-HERE, IF-CARRYING-FOOD, MOVE-TO-ADJACENT-FOOD-ELSE, MOVE-TO-ADJACENT-PHEREMONE-ELSE, and PROGN (a sequential instruction executor). The raw fitness of a program is defined as how many out of the 144 food pellets are transported to the nest within the allotted time of 400 simulation time steps. The genetic program is run for a maximum of 51 generations, with a population of 500 individual programs in each generation.

Koza reports in [7] that for this example, based upon seventeen runs of the experiment, processing 32 generations in each of three runs (equivalent to processing 49 500 individual programs) yields a 99% chance of creating a 100% fit individual (ie. program) for solving the problem. An example solution contained 47 points (ie. functions and terminals), hand-reducible to 18 points.

As a very simple example of the application of the BTIM, this learning task can be considered from the viewpoint of reducing uncertainty in the application domain. In particular, we can define a vector of random variables $X = X_1, X_2, \ldots, X_{144}$, where each $X_i \in X$ represents the position of a particular food pellet. There are 32 x 32 = 1024 different grid positions, each of which can be numbered such that each particular value x_i falls within the range from 1 to n = 1024, representing the particular grid position in which a particular food pellet is located. The simulation of central place food foraging begins with a random distribution of pellets across the grid. Hence all grid positions are equally probable locations of a particular pellet. For all x_i equally probable, $H(X_{i\text{-initial}}) = \ln n = \ln 1024 = 6.9315$ Nats. The value of 6.9315 Nats represents the amount of information gained from knowledge of the position of food pellet i at the beginning of the scenario. The trials run by Koza suggest a 99% probability of generating a successful program after three runs of 32 generations each. The successful program places all individuals within a 3 x 3 nest area, so reducing the possible values of X_i to a the range from 1 to 9. Applying the basic definition of $H(X_i)$, this results in an entropy after three runs of 32 generations per run of:

$$H(X_{i\text{-final}}) \quad = \quad -1015 \times (0.01 / 1015) \ln (0.01 / 1015)$$

$$- 9 \times (0.99 / 9) \ln (0.99 / 9)$$

$$= 2.3005 \text{ Nats.}$$

In this case, the uncertainty of the GP producing a result is taken into account in the calculation of the probabilities of the final position. The *adaptation* in relation to X_i is:

$$A(X_i) \quad = \quad H(X_{i\text{-initial}}) - H(X_{i\text{-final}}) \quad = \quad 4.6310 \text{ Nats}$$

Since this amount of adaptation is achieved for all of 144 food pellets, the total adaptation of the system is $A(X) = 144 \times 4.6310 = 666.9$ Nats. Koza defines fitness for this process as success within 400 simulation cycles of program execution. The *adaptation rate* for this problem can be defined either for the executing program, or for the genetic method of producing the program. The adaptation rate of the program can be taken as the adaptation achieved within the upper time bound of execution (400 cycles) used in the GP fitness measure. The adaptation rate is then:

$$\Delta A(X) \quad = \quad A(X)/\Delta t \quad = \quad 1.6672 \text{ Nats/cycle}$$

The successful program described by Koza in [7] contains 47 function and terminal points, and the program is represented in each of 20 simulated ants. Taking the total number of function and terminal points (in all ants) as a measure of the structural complexity of the solution, the *adaptive efficiency* of the system in relation to X is:

$$E_A(X) \quad = A(X)/C = 144 \text{ pellets} \times 4.631 \ / \ (47 \times 20) = 0.70947 \text{ Nats/atom}$$

A 99% chance of obtaining a successful individual requires three runs of the Genetic Program, for 32 generations per run. This amounts to 49 500 individuals (ie. candidate programs) that must be processed. On the conservative assumption that the average structural complexity of an individual is $47/2 = 23.5$ atoms, this amounts to a total generated complexity of 1 163 250 atoms. Based upon this figure, the adaptive efficiency of the Genetic Programming process for this problem (for the given parameter settings) is $666.9/1 163 250 = 5.7331$ E-4 Nats/atom.

7.3   Evolving Subsumption Architectures

This section again takes results reported by Koza in [7], and then analyses those results using the BTIM. Koza addresses learning a version of the subsumption architecture in which the layered competencies of the system each constitute a behaviour having an applicability predicate to turn the behaviour on or off, and an output feeding into a suppressor node at the output of the next lower level of the architecture. The hierarchical arrangement of suppressor nodes is taken as establishing a priority among behaviours. The applicability predicate of a behaviour is taken to be a composition of conditional logic functions and environmental input sensors (and possibly states of various alarm clock timers). The action part of each behaviour is taken to consist of a composition of functions that perform or initiate actions (typically affecting the environment or setting the internal state of some alarm clock timers). The hierarchical arrangement of suppressor nodes operating on the initial output actions of behaviours consists of a composition of logic functions, particularly IF-THEN-ELSE functions to determine whether an output is defined by a lower level behaviour, or by a higher level behaviour that suppresses the lower level. The evolution of this version of a subsumption architecture involves finding the appropriate layered behaviours, the applicability predicate and action set for each behaviour, and the appropriate conflict resolution herarchy.

The *wall following task* considered by Koza in [7] is performed by a robot that has twelve sonar sensors, each reporting the distance to the nearest wall as a floating-point number in feet. Each sensor covers 30°, so the total field of view is 360°. A single Boolean STOPPED sensor indicates if the robot has stopped moving. There are five primitive motor functions, including those for moving forward, moving backwards, stopping, turning right (by 30°), and turning left (by 30°). Three constant parameters respectively indicate the preferred following distance from the wall, a minimum safe distance, and a danger zone. Mataric's original design involved four LISP programs, STROLL, AVOID, ALIGN, and CORRECT, each corresponding to one task-achieving behaviour of the layered competency model of the robot. The applicability predicates of the behaviours were designed to be mutually exclusive, thereby eliminating the need for a conflict resolution strategy implemented via suppressor nodes. These LISP programs contained a total of 25 different atoms and 14 different functions.

Koza in [7] uses a simulated robot, so the danger zone, STOPPED sensor, and stop motion command are not needed. The simulated room is continuous (ie. not a grid). The room is a 27.6 foot square with convex irregularities protruding from two adjacent walls. The terminal set contains all of the remaining primitive motor commands, all of the remaining sensors and constants, and a derived value, SS, indicating the shortest returned sonar value. The function set is limited to a branching conditional, IFLTE, and a sequencing function, PROGN2. The moving and turning functions return the minimum value of the two most forward-directed sonar sensors of the robot. Raw fitness is defined as the proportion of the perimeter of the room traversed by the robot within a specified amount of time (400 time steps). The perimeter is envisaged as being defined by a series of 2.3 foot square tiles (this being the edging distance used by Mataric). There are 56 tiles in the complete perimeter. Success is defined as a traversal of each perimeter tile within the allowed 400 time steps. Raw fitness is defined as the actual number of tiles traversed within that time. At the start of a trial, the robot is located near the middle of the room.

The Genetic Program used a population size of 1000 and ran for a maximum of 101 generations. Koza reports in [7] that by generation 57, the best-of-generation individual, containing 151 points, attains a perfect score of 56. The S-expression can be simplified by hand to one of only 59 points, constituting an individual composed of conditional branching operations.

The perimeter tiles used by Koza to define success within the wall-following task can be used to define a grid across the room. In this case there are 132 grid positions within which the robot is randomly situated at the beginning of the task. Of these 132 positions, 56 are perimeter positions. For current purposes, the learning task can again be considered from the viewpoint of reducing uncertainty in the application domain. A random variable X represents the position of the robot. Each grid position is numbered such that each particular value x falls within the range from 1 to $n = 132$, representing the particular grid position in which the robot is located. Initially, all grid positions are equally probable locations of the robot, so, $H(X_{initial}) = \ln 132 = 4.8828$ Nats, representing the amount of information gained from knowledge of the position of the robot at the beginning of the scenario. After execution of the successful program, the robot is known to be in one of 56 perimeter positions, so the positional entropy of the robot is:

$$H(X_{final}) \quad = \quad -56 \times (1/56) \ln (1 / 56) \quad = \quad 4.025 \text{ Nats}$$

The *adaptation* is:

$$A(X) \quad = \quad H(X_{initial}) - H(X_{final}) \quad = \quad 0.8574 \text{ Nats}$$

The *adaptation rate* for the executing program is the adaptation achieved within the upper time bound of execution (400 cycles) used in the GP fitness measure (although Koza actually allocates this upper time bound for a *traversal* of the perimeter). The adaptation rate is then:

$$\Delta A(X) \quad = \quad A(X)/\Delta t \quad = \quad 0.0021 \text{ Nats/cycle}$$

The uncertainty of the genetic algorithm is not known in this case. The program that achieves this adaptation contains 145 function and terminal points. Again taking the number of function and terminal points as a measure of the structural complexity of the solution, the adaptive efficiency of the system in relation to X is:

$$E_A(X) \quad = A(X)/C = 0.8574 / 145 = \quad 0.0059 \text{ Nats/atom}$$

The solution program is found after processing 57000 individuals. On the conservative assumption that the average structural complexity of an individual is 145/2 = 72.5 atoms, the adaptive efficiency of the Genetic Programming process for this problem (for the given parameter settings) is $0.8574/(72.5 \times 57000) = 2.075$ E-6 Nats/atom.

The wall-following task appears very different from the emergent food foraging task. The respective BTIM metrics provide a basis for comparison of the performance of the Genetic Programming method in each of these cases. The process was found to have an efficiency of 5.7331×10^{-4} Nats/atom for the emergent food foraging process, and an efficiency of 2.075×10^{-6} Nats/atom for wall following. These figures can be equated with the overall merit of each system (ignoring coordination rates, F_c, and time to produce the respective solutions) indicating an apparent disparity in the merit of GP in these tasks on the order of two magnitudes. The explanation for this is that the measured goal state of the wall-following task was simply to place the robot within the perimeter, whereas the real goal in the problem as formulated by Koza as reflected in the fitness function was to traverse the perimeter. Hence, the low merit of the system that generates the wall following solution can be interpreted as indicating either the generation of too complex a solution for the problem (assuming that the goal is to reach the perimeter), or too simplistic an interpretation of the goal state (which should be that of traversing the perimeter). The adaptation rates of the two problems (1.6672 Nats/cycle for emergent food foraging and 0.0021 Nats/cycle for wall-following) clearly show the differences in merit of the two systems, highlighting the loss of merit that arises when the fitness function is inappropriate for the measured goal.

8 Conclusion

Valavanis and Saridis [7] have proposed a Theory of Intelligent Machines based upon a formulation of the principle of Increasing Precision with Decreasing Intelligence justified in terms of the constancy of the entropy rate in the performance of a task by an intelligent system. This paper has shown how the hierarchical model developed by Valavanis and Saridis is not the only model of intelligent system architectures that is compatible with the use of general systems-theoretic analysis methods. In fact, Conant's general systems theory (see [4]) suggests desirable features of intelligent systems that are more characteristic on non-hierarchical, behavioural approaches to intelligent agent design. A novel formulation of a Behavioural Theory of Intelligent Machines (BTIM) has been presented, based upon concepts from general systems theory. The BTIM captures the performance and design characteristics that typify behavioural systems and expresses those characteristics in terms of a program to maximise a quantitative metric describing the merit of a system. However, the BTIM as defined above has several limitations:

- it cannot be applied unless the behaviour of system variables can be described probabilistically. In particular, a system of high merit is one that increases the probability over time of achieving system goals, expressed in terms of system variables.

- the notion of an information *channel* is not always easy to identify, and in fact only figures in the merit definition given above as part of the coordination rate of the system. A channel is an information path by which information about one system variable may provide information about another system variable, thereby reducing the entropy of the latter. The human "owner" of a system is generally interested in whether the system satisfies goals valued by the owner. However, those goals may not be expressible in terms of system variables, or in terms of transmission between connected system variables. The applicability of information-theoretical concepts may therefore be questioned.

- a valid method of measuring adaptation is needed. The method should account for the minimum time needed for specific instances of adaptation, the need for constant readaptation between the time limits due to dynamic circumstances, and changing (and possibly nonlinear) adaptation characteristics beyond the scope of the measurement period.

The BTIM has been demonstrated by an analysis of the performance of Genetic Programming in generating two different behavioural system designs. The Behavioural Theory of Intelligent Machines provides a powerful basis for the analysis, design, and evaluation of autonomous agents, within the limitations noted. Importantly, the metrics that BTIM provides can be used for a more "objective" comparative analysis of different agent designs, and of behavioural designs against designs based upon traditional representation-centred methods. Many hypotheses of the behavioural paradigm can be measured and tested using information-theoretic metrics, including the hypotheses that behavioural systems are more robust, more flexible, more reliable, more adaptive, and simpler than representation-centred architectures.

References

1. Brooks, R. A.: A Robust Layered Control System For A Mobile Robot. IEEE Journal of Robotics and Automation, RA-2 1 (1986) 14-23.

2. Brooks, R. A.: Challenges for Complete Creature Architectures. Meyer J.-A. and Wilson S. W. (eds): From Animals to Animats: Proceedings of the First International Conference on Simulation of Adaptive Behaviour , MIT Press (1991) 434-443.

3. Conant, R. C.: Detecting Subsystems of a Complex System. IEEE Transactions on Systems, Man, and Cybernetics (1972) 550 - 553.

4. Conant, R. C.: Laws of Information which Govern Systems. IEEE Transactions on Systems, Man, and Cybernetics SMC-6(4) (1976) 240 - 255.

5. Georgeff, M. P.: Planning. Annual Reviews in Computing Science (1987) 359-400.

6. Holland, J. H.: Adaptation in Natural and Artificial Systems, The MIT Press (1992).

7. Koza, J. R.: Genetic Programming, The MIT Press 1992.

8. Lindley, C. A.: Extending the Behavioural Paradigm for Intelligent Systems. Twenty-Seventh Annual Hawaii International Conference on System Sciences, Maui, HI (1994).

9. Maes, P.: Behaviour-Based Artificial Intelligence. Proceedings of the Second International Conference on Simulation of Adaptive Behaviour MIT Press (1993).

10. Valavanis, K. P., Saridis, G. N.: Intelligent Robotic Systems: Theory, Design and Applications, Kluwer Academic Publishers (1992).

On Evolving Robust Strategies for Iterated Prisoner's Dilemma

P. J. Darwen and X. Yao

Department of Computer Science
University College, The University of New South Wales
Australian Defence Force Academy
Canberra, ACT, Australia 2600

Abstract. Evolution is a fundamental form of adaptation in a dynamic and complex environment. Genetic algorithms are an effective tool in the empirical study of evolution. This paper follows Axelrod's work [2] in using the genetic algorithm to evolve strategies for playing the game of Iterated Prisoner's Dilemma, using co-evolution, where each member of the population (each strategy) is evaluated by how it performs against the other members of the current population. This creates a dynamic environment in which the algorithm is optimising to a moving target instead of the usual evaluation against some fixed set of strategies. The hope is that this will stimulate an "arms race" of innovation [3].

We conduct two sets of experiments. The first set investigates what conditions evolve the best strategies. The second set studies the robustness of the strategies thus evolved, that is, are the strategies useful only in the round robin of its population or are they effective against a wide variety of opponents?

Our results indicate that the population has nearly always converged by about 250 generations, by which time the bias in the population has almost always stabilised at 85%. Our results confirm that cooperation almost always becomes the dominant strategy [1, 2]. We can also confirm that seeding the population with expert strategies is best done in small amounts so as to leave the initial population with plenty of genetic diversity [7].

The lack of robustness in strategies produced in the round robin evaluation is demonstrated by some examples of a population of naïve cooperators being exploited by a defect-first strategy. This causes a sudden but ephemeral decline in the population's average score, but it recovers when less naïve cooperators emerge and do well against the exploiting strategies. This example of runaway evolution is brought back to reality by a suitable mutation, reminiscent of punctuated equilibria [12]. We find that a way to reduce such naïvity is to make the GA population play against an extra, static, high-quality strategy (not part of the GA population), as well as all the rest of the population. The strategies thus produced perform better against opponents that were included in the round robin (as expected) and, more significantly, better against opponents that were *not* included. That is, robustness is improved.

1 Introduction

Evolution works. In nature, evolution produces increasingly fit organisms in uncertain environments. Evolution is a proven form of adaptation to deal with a dynamic and complex environment. Emulating evolution shows great promise for machine learning.

The genetic algorithm (GA) [10] is a search algorithm that emulates some basic features of natural evolution. The GA works well in searching spaces with the same difficulties faced by machine learning when searching the space of possible strategies in games of conflict [4].

We use the GA to learn to play the iterated prisoner's dilemma (IPD) [1], a simple two-player non-zero-sum game which is widely studied in such diverse fields as machine learning, economics, political science, and mathematical game theory.

1.1 Background

Axelrod [1] investigated evolution and the IPD, and simulated a population (of strategies) in which each strategy plays IPD with every other individual [2, p 38]. In such a system, the environment (made up by the members of the evolving population) is constantly changing. He found that this dynamic environment produced strategies that performed very well against their population. Fogel [5] studied an almost identical system.

This begs a question: can a strategy evolved from competing against the its own population succeed against strategies not in its own population? Drawing an analogy from biology, can a successful predator on one continent be effective against the successful predators of some other, isolated continent? In machine learning, we want to produce strategies that are effective against all possible opponents, not just those in the local population. This desirable quality is "robustness".

Lindgren provided a radical insight [12]. He simulated a system similar to those of Axelrod [2] and Fogel [5]. He showed that (under certain conditions) when a population (of strategies) plays IPD against its own members, the high-performing strategies who dominate the population for long periods of time, are suddenly wiped out and replaced. The results of Lindgren's simulation bear an uncanny resemblance to the "punctuated equilibria" of natural evolution [3]. "In particular, the large extinctions that appear in these simulations should be studied in more detail, since these collapses are triggered by the dynamical system itself and do not need external catastrophes for their explanation." [12, p 310].

From the point of view of machine learning, Lindgren [12] demonstrated that co-evolution produced strategies that were not robust, i.e., the strategies did well against the local population, but when something new and innovative appeared (created by the the evolutionary process) they failed dismally. Fogel [5] also noticed that co-evolved strategies could do well against each other, but still have big flaws that could be exploited by the right opponent.

Why did they seemingly-invincible strategies fail? How can we prevent this disaster happening to strategies produced by evolutionary machine learning? This paper answers these questions.

1.2 The Answers

We study a system similar Axelrod's [2], and find a reason for Lindgren's catastrophic collapses [12].

It turns out that when a near-expert strategy dominates a population, the homogeneity in their behaviour (in the case of IPD, they all cooperate) means certain features atrophy. In the case of IPD, tit-for-tat decays; in a cooperative environment, it can pay to *not* retaliate, and this atrophy of a once-useful reaction opens the way for exploitation.

In a fanciful analogy, imagine a species of big-horned buffalo that must compete with other species of buffalo. Our species' huge horns make other lesser-horned species extinct. The now-dominant big-horned species then finds that their big horns are a liability (extra weight, etc.) and evolution selects for smaller horns. The now small-horned species dominates until a random mutation (or migration from elsewhere) allows a big-horned sub-species to exterminate their now-small-horned opponents.

In our paper, this drama occurs between the players of iterated prisoner's dilemma. This sheds a light on the robustness problem in strategies produced by co-evolution, and on evolution in nature.

2 Genetic algorithms

2.1 Introduction

Machine learning often involves searching a space of legal alternatives for near-optimal solutions [4, page 614]. The genetic algorithm (GA) is a powerful technique for searching spaces that suffer the following difficulties:-

1. The search space is combinatorially large;
2. Little is known *a priori* about the search space;
3. The search space contains fine-grained discontinuities, local optima, and other irregularities.

Games whose strategies involve identifying and exploiting an opponent are search spaces of this character [4]. Hence, the GA seems a suitable machine learning technique to learn to play iterated prisoner's dilemma (IPD).

2.2 Overview of the Genetic Algorithm

A genetic algorithm (GA) [10] maintains a population of sample points from the search space. It represents a point by a string of (usually binary) characters, known as a *genotype*.

For example, a common way to represent points over a continuous parameter $x : 0 \leq x < 1$ is to discretize the space into (say) 1024 points, and represent a point by a binary string of 10 bits (where $2^{10} = 1024$). Point 776 of the 1024 possible points would be represented by the binary string 0110000101 .

A GA begins with a population of these binary strings, usually initialised with random contents, and at each generation, the following happens:-

Evaluation: Each member of the population is evaluated according to the function to be optimised.

Selection: The better performers of the population are selected for reproduction, i.e., the worse performers are erased.

Reproduction: The genetic material of the better performers is made into a new population using various genetic operators, including crossover (emulating sexual reproduction), mutation, and numerous variations [14].

Mutation is changing a character in a genotype (from 0 to 1 or vica versa) in a new individual. The mutation rate is usually low; 1 mutation per 1000 bits copied is normal [6, p 14].

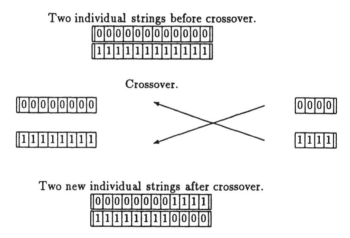

Fig. 1. Crossover is when genetic material is swapped and recombined, creating two different individuals.

Crossover is the dominant mechanism of genetic rearrangement for both real organisms and genetic algorithms [11]. The simplest implementation of crossover is to pick two genotypes (usually strings of characters), randomly choose a common crossover point, and cut and paste to create two new genotypes, as in Figure 1.

2.3 Why it works

The genetic algorithm quickly locates high-payoff "target" regions of the search space because each single string samples many regions. For example (where "*" indicates that a bit's value is unspecified), this string:-

| 1 | 0 | 0 | 1 | 1 | 0 | 1 | 1 | 0 | 0 | 0 |

is a member of this region (one quarter of the search space):-

| 1 | 0 | * | * | * | * | * | * | * | * | * |

as well as this region ($\frac{1}{32}$ of the search space):-

| * | 0 | 0 | * | * | 0 | * | * | 0 | 0 | 0 |

and so on. In general, if the binary strings are of length L, there are

$$\binom{L}{N}$$

distinct N-th order partitions of the the space (where an nth-order partition divides the space into 2^n parts) [4]. This implicit parallelism allows a genetic algorithm's population to sample many more regions than its size.

A disadvantage: while a GA can rapidly find near-optimal solutions, other methods are better at going from near-optimal to optimal [11, 7]. An example from natural evolution is the giraffe: the giraffe's larynx (or voice box) is near its brain, and an optimal giraffe would have a short brain-to-larynx nerve. In reality, this nerve runs from the brain all the way down the giraffe's neck, loops around a major blood vessel, then runs back up the neck to the larynx. Like natural evolution, genetic algorithms usually give a near-optimum, but not always the optimum.

3 Strategy Generation in IPD

We use the GA to generate expert strategies to play the simple game of iterated prisoner's dilemma (IPD).

3.1 The Game

In Prisoner's Dilemma, each player can either cooperate with the other player, or defect. In iterated prisoner's dilemma (IPD), this step is repeated many times, and each player can remember previous steps. We used Axelrod's variant of prisoner's dilemma [2], is shown in Figure 2.

The number of rounds per game is fixed at 50 rounds; that is, each pair of opponents play the game in Figure 2 for 50 rounds, remembering the previous outcomes.

	Cooperate	Defect
Cooperate	3 3	5 0
Defect	0 5	1 1

Fig. 2. Axelrod's variant [2] of Prisoner's Dilemma as used in the runs described below.

3.2 Software Implementation

Our implementation follows Axelrod [2]. Lindgren [12] used a variation of Axelrod's approach, and Fogel [5] used evolutionary programming which is strictly speaking not a GA. However, the three approaches are broadly similar, and our results are relevant to all of evolutionary machine learning.

Several excellent public-domain genetic algorithm packages are available. We used Grefenstette's GENESIS 5.0, and modified the evaluation procedure to evaluate individuals from how they score against other members of the population.

Genotype We used the same genotype as Axelrod [2] Only the three most recent steps are remembered. Since each step has 4 possible outcomes (see Figure 2), that means that there are $4 \times 4 \times 4 = 64$ possible histories of 3 steps. The genotype lists an action (cooperate or defect) to take for each of the possible 64 histories. These 64 actions are the parameters that the GA varies.

For example, if my previous 3 actions were cooperate-defect-cooperate (010), and my opponent's last 3 actions were defect-cooperate-cooperate (100), then we juxtapose these (opponent's history first) to get $100010 = 34$, and take the thirty-fourth action in my genotype. My opponent juxtaposes these to get $010100 = 20$, and does the action in the twentieth position of his genotype.

For the first 3 steps of a game of IPD, there are less than 3 previous steps to look back to, so the actions to take for these first 3 steps must also be specified in the genotype. The initial conditions are specified by an additional 6 bits, which makes 70 bits in the genotype. These extra 6 bits carry the assumed (pre-game) 3 steps of both players, so that the extra 6 bits alone indicate which of the (original) 64 actions to take on the first round. For rounds 2 and 3, the actual results of the game are updated to the history, so the history (as seen by an individual) contains both the pre-game "history" (carried by the genotype's extra 6 bits) as well as the actual events. In step 4 and afterwards, only the actual results are used.

The Round Robin An individual strategy is evaluated by playing it against all the other members of the population. For a population of n individuals, the number of games to be played is $\sum_{i=1}^{n} i = (n-1)\frac{n}{2}$ if an individual plays against itself, and one less if it does not. The sum of the scores that an individual achieves in all these games is its fitness.

4 Results on Co-evolution

4.1 The Evolution of Cooperation

An interesting feature of evolution and iterated prisoner's dilemma is the evolution of cooperation. Even though the we simulate the most mercenary, dog-eat-dog form of evolution, cooperative strategies eventually proliferate and drive their non-cooperative competitors to extinction. Axelrod [1] describes this phenomenon in great detail.

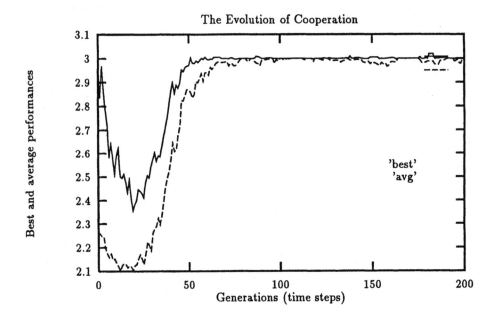

Fig. 3. The evolution of cooperation in iterated prisoner's dilemma, starting from a random population. At first, non-cooperative strategies do well and performance plummets as non-cooperative strategies proliferate and they run out of victims. Eventually, cooperative strategies dominate and the performance rises to the level caused by mutual cooperation.

Figure 3 demonstrates the evolution of cooperation. From an initial population of random strategies, the population's average payoff plummets as non-cooperative strategies exploit more naïve genotypes. After a few dozen generations, gullible victims become extinct, and the population fills with strategies that cooperate when they can and retaliate against non-cooperative strategies, who are driven to extinction. In the end, only cooperative strategies survive and average per-step payoff indicates mutual cooperation.

4.2   Experiment on length of run

How long (if ever) does it take for the genetic algorithm to converge? That is, we are expecting evolution to usually produce a population of high-performance strategies that, once achieved, will change very little. Figure reffig:evcoop indicates cooperation dominates after about 50 generations. Does further convergence take place after this? This experiment is simply to find how long the genetic algorithm usually runs until it stops evolving, starting from a random initial population.

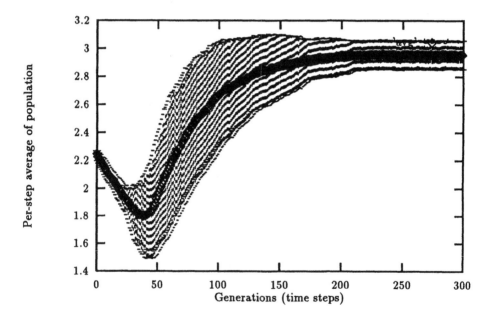

Fig. 4. This shows the population's average per-step performance, and standard deviation, taken over 30 different runs. This indicates that, on average, the system has converged after about 250 generations.

We did thirty runs, each starting from a (different) random initial population of 100 individuals. The round robin was the same as Axelrod's [2]; all members of the population compete against every member of the population, *including* itself.

Figure 4 shows the average score of the GA population. Figure 5 shows the mean and standard deviation of the bias[1] of the population. The 30 runs indicate that 250 generations is a satisfactory run length.

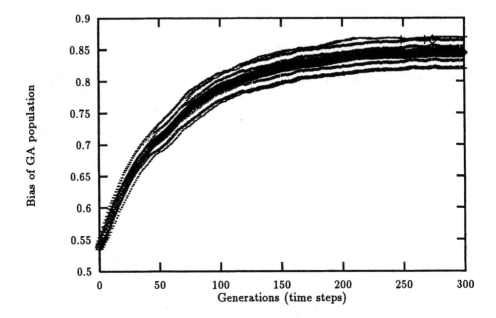

Fig. 5. The bias of a genetic algorithm's population is the average per-bit level of convergence in the population; for example, a bias of 0.75 means that, on average, each position has converged to 75% zeros or 75% ones. This graph shows the mean and standard deviation of the bias for 30 runs of a genetic algorithm population playing repeated prisoner's dilemma in a round robin. The system has converged by 250 generations.

By 250 generations, the bias in the population has stabilised (Figure 5) at 85%. Why not 100%? If there is still diversity in the population, why don't the better performers multiply and the worse performers die out?

Recall that the genotype lists actions to take under all contingencies. Diversity in the population means that individuals would choose different actions for the same contingency. We may infer that the variation in the population means this difference in opinion carries no benefit or penalty, and that some contingencies (represented by those diverse opinions) rarely arise in a cooperating population.

---

[1] The *bias* of the population of a genetic algorithm is the average percentage of the most prominent value in each bit position. That is, a bias of 0.75 means that, on average, each position has converged to 75% zeros or 75% ones. The minimum bias is 0.5, the maximum is 1.

An analogy: imagine a species which lives where there are no flowering plants, and whose genes (by accident) make some individuals allergic to flowering plants. Since there are no flowering plants, this diversity carries no penalty or reward, and this diversity can continue despite the evolutionary pressure to succeed. Should flowering plants appear, though, selection will operate and quickly select for the better performers.

In our simulation, if no "flowering plants" appear then this diversity can continue to exist.

4.3 Experiment on seeding the initial population

One improvement to a round-robin genetic algorithm (i.e., a GA where the fitness of an individual is found from how it competes with the other individuals of the evolving population) is being investigated by Grefenstette [13]: seed the initial population with strategies known to be good.

We will find the best initial population that consists of a mix of a random initial population and a known effective strategy, tit-for-tat.

Each run used a GA population of 100 individuals. The proportion of the initial population that was tit-for-tat was varied from 0% to 25% in steps of 1%, and then from 30% to 70% in steps of 10%. For each mix of tit-for-tat and random strategies in the initial population, 30 runs were done (each of 250 generations), each using a different random component of the initial population.

Figure 6 shows how the average strategy of the final population scored against both tit-for-tat (where it nearly always scored 3, indicating mutual cooperation) and against a large number (> 1000)of random genotypes. For each run, the average genotype of the final population was played against both tit-for-tat and a large number of random genotypes, giving the mean and standard deviation for the 30 for each mix of the initial population. This is shown in Figure 6.

As we expect from the evolution of cooperation in a co-evolving population [1], almost all final strategies do equally well against tit-for-tat.

Surprisingly, the best results against random strategies are *not* achieved when the initial population is all random, as one might expect. The best strategies are produced when about 10% of the initial population is tit-for-tat, and the rest random, as shown in Figure 6. This indicates that genetic diversity produces better strategies, outweighing the effect of the environment matching the task. That genetic diversity aids performance agrees with Grefenstette [7].

This indicates that seeding an initial population is a worthwhile idea, but genetic diversity in a GA is still more important.

5 Results on Robustness

5.1 Glitches and the robustness problem

There is a problem with strategies produced in a closed population of competing strategies: are the produced strategies good against all possible opponents, or are they useful only against the local population?

Performance of average genotype of final population starting from a mix of random and Tit-for-tat

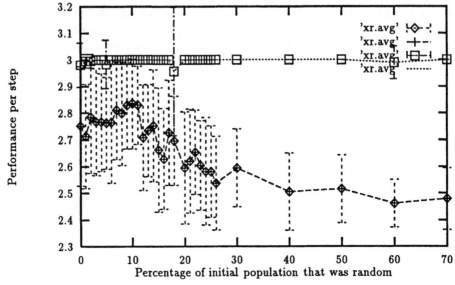

Fig. 6. This shows the results of the final population against both tit-for-tat and a large number of random rules, when the initial population contains tit-for-tat. The best results are obtained when around 10% of the initial population is tit-for-tat.

In Figure 7, we see that after settling down to cooperation (similar to Figure 3), in generation 222 a new mutation arises which successfully exploits its cooperative neighbours. This successful uncooperative strategy proliferates until there are so many uncooperative members of the population (and thus so few cooperative victims) that it becomes an unprofitable strategy, and cooperation returns.

The Cause of the Glitch What causes the "glitch" in Figure 7? Recall that the genotype of an individual is simply a look-up table that lists 64 actions (either "cooperate" or "defect"), corresponding to the 64 possible histories of the most recent 3 steps.

Consider this case: following a time of mutual cooperation, your opponent defected on you. Your opponent's last three actions are now defect-cooperate-cooperate, and your last three actions are cooperate-cooperate-cooperate. In the genotype's scheme (where "0" represents cooperation and "1" represents defection) this particular history of the last 3 steps (100-000, binary for 32) tells us to do the action listed in position 32 of our genotype.

Thus, this out-of-the-blue defection would cause the strategy to take the action (cooperate or defect) listed in bit 32 in the look-up table that forms the genotype. A strategy of tit-for-tat would have this action listed as a defection (1), to retaliate. But what happens in a closed population of cooperators?

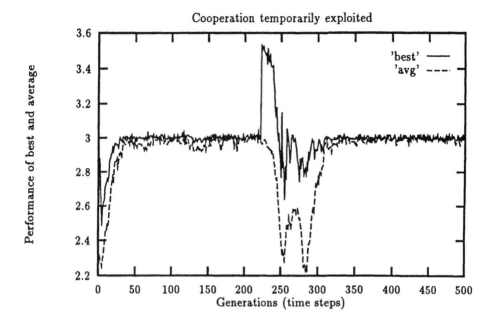

Fig. 7. This shows that a new and innovative mutation can exploit a population that has cooperated for so long that they can be suckered.

In a population of cooperators, unilateral defection rarely happens, and one of two things may happen to the part of the genotype (bit 32) that deals with this contingency:-

Genetic drift: There may be genetic drift: that is, since there is no selection pressure one way or the other on this part of the genotype, random mutations will fill this part of the genotype with garbage, since its contents will make no difference to fitness. Selection for retaliation or cooperation after a unilateral defection will not occur if you never have a unilateral defection.

Selection: The method that previously worked is selected against. In our example (Figure 8), first there is selective pressure on bit 32 (the action to take after a unilateral defection) to retaliate; but after generation 130, there is selective pressure to *cooperate* after a unilateral defection. This is because in this particular local population dominated by cooperation, it pays to do this.

In Figure 7, most individuals had a 0 (cooperate) in bit 32 by generation 200, which was an advantage in this particular cooperative population. Random mutation hit upon a strategy that could exploit this particular flaw, which exploited this over-cooperative flaw. This is shown by Figure 8: before the successful exploitative mutation at generation 222, most of the population contained a 0 in bit 32, and would cooperate after suffering a unilateral defection. This was the flaw exploited by the new mutation.

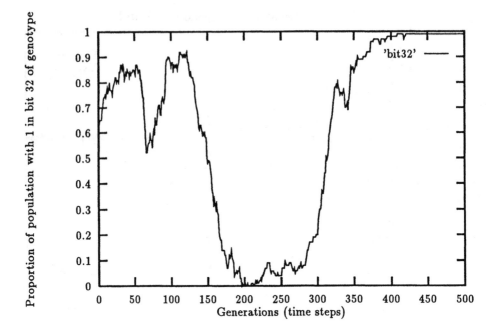

Fig. 8. This shows that a new and innovative mutation can exploit a population that has cooperated for so long that bit 32 (the one that contains the action to take after a unilateral defection) is 0 (cooperate) for most of the population.

A fanciful analogy may help explain this: imagine a species of big-horned buffalo competing with other species of buffalo. Our species' huge horns make lesser-horned species extinct. The now-dominant big-horned species finds their horns are a liability (extra weight to carry, etc.) and evolution selects for smaller horns. The now small-horned species dominates until a random mutation (or migration from elsewhere) allows a big-horned sub-species to exterminate their small-horned opponents. In Figure 8, put "horn size" on the y-axis and a similar drama occurs.

Another "glitch" is shown in Figure 9. Lindgren [12], in a similar simulation, demonstrated that these events can happen repeatedly if the simulation runs for long enough.

We want to avoid disastrously naïve situations like this. We investigate two possible ways to avoid this kind of glitch:-

- Seed the initial population with some known expert rules, instead of a completely random initial population.
- Include a few extra, unchanging genotypes in the round robin, so that each individual in the population has to play every other member, plus the extra players. In the analogy of the old Western town, this is like having a resident bad man, so that people will know what to do in that contingency.

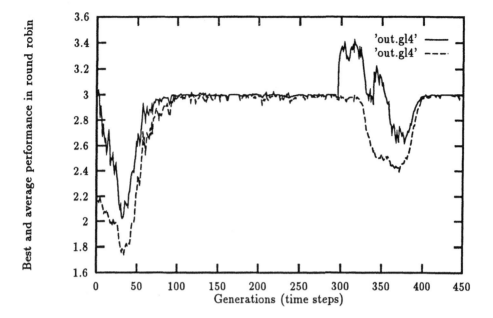

Fig. 9. Another example of an innovative mutation that exploits a population which has cooperated for so long that they can be suckered.

5.2 Experiment on static rules in round robin

We have seen how a strategy that evolved in a co-evolving population runs the risk of failing badly against new opponents. That is, there is a risk of blithely generating naïve strategies that are good against similar strategies but vulnerable to radically different ones. This situation is described in the previous section, is shown in Figure 7, and also occurred in Fogel's similar approach evolutionary learning to play iterated prisoner's dilemma [5].

One solution is to expand the round robin in the genetic algorithm; as well as the other members of the evolving population, each individual could also play against extra, static strategies. The aim is to prevent over-specialisation to the local population, and produce versatile strategies that work against all opponents.

In this experiment, we add to the round robin. We do 4 sets of 100 runs (each of 250 generations; 100000 generations altogether) where the round robin contains:-

1. The usual round robin, with only the evolving population.
2. As for 1, but including a random genotype. This is a genotype of 70 bits just like the other members of the population, but its contents are random, and stay the same for a generation, before being replaced by a different random genotype in the next generation.
3. As for 1, but including the expert strategy tit-for-tat [1].

4. As for 1, but including both a (different) random strategy each generation, and the expert strategy tit-for-tat [1].

How can we evaluate the contents of a final population? That is, how do we tell if one round robin learned to play IPD better than another?

For each of the 400 runs, we followed the following procedure:-

1. Run for 250 generations.
2. From the final population (of 100 individuals), find the *average* strategy. That is, if a particular bit in the genotype is "0" for 50% or more of the population, it is 0 in the average genotype.
3. We evaluate this average strategy by running it against two strategies:-

   Tit-for-tat This strategy is to verify that the strategy does cooperate in repeated prisoner's dilemma against an expert opponent.

   Many random strategies We also play a rule set against a large number (> 1000) of random genotypes; that is, look-up tables filled with random contents. This is to verify that the strategy can exploit non-expert opponents.

Table 1 shows the average and standard deviation of the scores against tit-for-tat and the random genotypes. The evolution of cooperation causes the population to cooperate, so the average genotype usually scores 3 against tit-for-tat, and a few spectacular failures brought the average down in table 1.

Against the exploitable random genotypes, the populations which faced a random opponent in the round robin did better; as expected, the closer the learning situation is to the real thing, the better the results.

In round robin	Against randoms	Against tit-for-tat
Population	2.54 ± 0.27	2.72 ± 0.47
Plus random	2.78 ± 0.23	2.66 ± 0.51
Plus tit-for-tat	2.55 ± 0.25	2.81 ± 0.40
Plus both	2.74 ± 0.23	2.72 ± 0.50

Table 1. The round robin of a genetic algorithm has extra rules added. For each round robin, 100 runs were done starting from a different random initial population each time. The average genotype of the final population scores best against an opponent that was added to the round robin.

A genetic algorithm will specialise to fit the problem presented [8]. If we include (say) tit-for-tat in the round robin, we expect the strategy produced to do better against tit-for-tat; table 1 agrees. This result also agrees with Grefenstette [9], where training under variable conditions created rule sets that performed well under variable conditions.

As a method of overcoming the robustness problem in co-evolutionary machine learning, adding static strategies seems to work. Table 1 shows that adding a fixed strategy to the round robin causes the population to succeed against *those* opponents.

Sometimes, we don't have the luxury of knowing the superior opponents *a priori*. The open question is still: how can co-evolution produce strategies that are robust performers against all possible opponents? That is, can a predator that evolved on one continent ever be sure of succeeding against a predator from another continent?

6 Discussion and Conclusion

Lindgren [12] and Fogel [5] noticed that strategies produced by the co-evolution of closed populations may do well in that closed population, but can fail against outsiders. we have demonstrated a possible cause of this lack of robustness: the homogeneity of some high-performing populations leads to the atrophy of techniques (such as retaliation against unilateral defection), opening the way for their exploitation. This is shown in "glitches" (see Figure 7, where a homogeneous GA population that was doing well in its own predictable population was exterminated by a new, innovative mutation, which exploited the lack of retaliation against unilateral defection, as shown in Figure 8.

This naïve vulnerability to innovation is the robustness problem in co-evolving genetic algorithms.

Adding static opponents to the round robin improves the results of the final population, as shown in table 1. As expected, the rules produced by a population do better against an opponent that was added to the round robin.

An open question is: can extra, fixed opponents produce strategies that are also good against opponents never seen before?

The results show that this method has great promise for automatic strategy generation.

References

1. Robert M. Axelrod. *The Evolution of Cooperation*. Basic Books, 1984.
2. Robert M. Axelrod. The evolution of strategies in the iterated prisoner's dilemma. In Lawrence Davis, editor, *Genetic Algorithms and Simulated Annealing*, chapter 3, pages 32–41. Morgan Kaufmann, 1987.
3. Richard Dawkins. *The Blind Watchmaker*. Longman, first edition, 1986.
4. Kenneth A. Dejong. Genetic-algorithm-based learning. In Y Kodratoff and R Michalski, editors, *Machine Learning*, chapter 21, pages 611–638. Morgan Kaufmann, 1990.
5. David B. Fogel. Evolving behaviours in the iterated prisoner's dilemma. *Evolutionary Computation*, 1(1):77–97, 1993.
6. David E. Goldberg. *Genetic Algorithms in Search, Optimization, and Machine Learning*. Addison-Wesley, 1989.
7. John J. Grefenstette. Incorporating problem specific knowledge into genetic algorithms. In Lawrence Davis, editor, *Genetic Algorithms and Simulated Annealing*, chapter 4, pages 42–60. Morgan Kaufmann, 1987.

8. John J. Grefenstette. Strategy acquisition with genetic algorithms. In Lawrence Davis, editor, *Handbook of Genetic Algorithms*, chapter 14, pages 186–201. Von Nostrand Reinhold, 115 Fifth Avenue, New York NY 10003, 1991.

9. John J. Grefenstette, Connie L. Ramsey, and Alan C. Schultz. Learning sequential decision rules using simulation models and competition. *Machine Learning*, 5:355–381, October 1990.

10. John H. Holland. *Adaptation in Natural and Artificial Systems*. MIT Press, second edition, 1992.

11. John H. Holland. Genetic algorithms. *Scientific American*, 267(4):44–50, July 1992.

12. Kristian Lindgren. *Evolutionary Phenomena in Simple Dynamics*, volume 10 of *Santa Fe Institute Studies in the Sciences of Complexity*, pages 295–312. Addison-Wesley, 1991.

13. Alan C. Schultz and John J. Grefenstette. Improving tactical plans with genetic algorithms. In *Proceedings of IEEE Conferenece on Tools for Artificial Intelligence*, pages 328–334. IEEE Society Press, November 1990.

14. Xin Yao. An empirical study of genetic operators in genetic algorithms. *Microprocessing and Microprogramming*, 38:707–714, 1993.

Comparison of Heuristic Search Algorithms for Single Machine Scheduling Problems

Graham McMahon and Daniel Hadinoto

School of Information Technology, Bond University
Gold Coast, Qld 4229, Australia

Abstract. This paper compares the performance of four heuristic search algorithms for single machine scheduling problems: local search, simulated annealing, tabu search and genetic algorithms. To investigate their potential, the algorithms are applied to a single machine scheduling problem to minimise tardiness of all jobs with arbitrary ready times, processing times and due times. This problem is known to be NP complete. The purpose of the comparison is to find a good approximation algorithm ie. the algorithm is not designed to search for an optimal solution to the problem.

1 Introduction

The purpose of this paper is to compare four heuristic search algorithms: local search, simulated annealing, tabu search and genetic algorithm. Each algorithm can be implemented in a variety of ways using various data structures and parameters. It would be a substantial task to explore all these possibilities. The aim of this paper is less ambitious. It is not to compare the best possible algorithms for each category which is very difficult to establish. Instead, a relatively simple implementation of each algorithm of interest is chosen. It is possible that further developments in any of these areas may affect the results of the comparison.

The concept of local search is based on the idea of repeatedly moving from one solution to another solution in its neighbourhood using a descent strategy (always moves in a direction of improvement). Local search, which is also called neighbourhood search, often results in a convergence to a local rather than a global optimum

Simulated annealing and tabu search, which can be considered as variants of the local search, attempt to overcome this problem of local optimality by accepting, in a limited way, transitions which correspond to an increase in the cost value. The effectiveness of these two search methods are, then, compared to genetic algorithms. Genetic algorithms are a randomised search technique which simulate the process of natural evolution. The algorithm maintains a population of solutions which evolve as solutions are created and discarded. Genetic operators are used to generate new solutions from the current population.

To investigate the effectiveness of these algorithms, they are applied to a single machine scheduling problem where the goal is to minimise the maximum tardiness. This problem is known to be NP complete (French, 1982). Therefore, it is unlikely that this problem has a polynomial time algorithm.

2 Problem Definition

Suppose there are n independent jobs to be scheduled on one machine. Each permutation of the n jobs defines a feasible schedule. The total number of solutions to this problem is equal to the number of different permutations of n which is n!. This type of scheduling problem is called a permutation schedule (Baker, 1974).

Each job has a ready time, a processing time and a due time associated with it. A job can not be processed before its ready time. The processing time is the duration of time the job must be processed by the machine. The due time is the time at which the job is to be completed. If the processing of a job is completed after its due time, then the job is considered to be tardy. The tardiness value of the job is obtained by subtracting its due time from its completion time. A job which is completed on or before its due time has a tardiness value of 0.

The performance measure used to evaluate a schedule is the maximum tardiness value. The problem is a minimisation problem ie. to minimise the maximum tardiness values of all jobs in the schedule.

3 Test Problems

Ten test problems were created with 100 jobs to be scheduled. The ready times of the jobs were generated randomly between 0 and 999 and their processing times were generated randomly between 1 and 20. The due times of the jobs were generated randomly between (ready time + 0) and (ready time + 99). The algorithms were implemented using the C programming language on a SUN SPARC IPX 25 mips.

4 Local Search

4.1 Neighbourhood Structure

The most important issue to be resolved when implementing local search method is the definition of the neighbourhood structure. A neighbourhood operator is needed to transform a solution into its neighbour solution. Two operators were evaluated and compared in this study: exchange operator (interchange two arbitrary jobs) and move operator (move an arbitrary job before another job)

The local optimisation algorithm employs two variations in moving from one solution to its neighbour solution:

1. Fastest descent: a solution is immediately replaced by its neighbour solution with a lower cost

2. Steepest descent: explores all its neighbour solutions before replacing it with the lowest cost neighbour

4.2 Preliminary Results

An experiment was conducted to compare the performance of the two operators using the two variations in moving from one solution to its neighbour solution. Each method started with the same initial random solution and terminated when 100,000 solutions had been generated. When the algorithm reached a local optimum, it restarts from another randomly generated solution. The maximum tardiness values (in time units) obtained at the end of each run is summarised in the table below.

Methods	P1	P2	P3	P4	P5	P6	P7	P8	P9	P10
Steepest/exchange	1004	841	794	1177	1042	874	1012	906	836	1094
Steepest/move	1252	1381	1511	1354	1375	1320	1534	1263	1528	1417
Fastest/exchange	1266	1334	1166	1298	1056	1204	1364	1326	1409	1246
Fastest/move	1517	1395	1400	1394	1367	1356	1422	1459	1447	1454

From this experiment we can see that the combination of the exchange operator and the steepest descent method is superior to the other methods ie. the best in all 10 problems. The initial solutions for each of these 10 problems were in the range of 1700 and 1800, while the global optimum values were in the range of 0 and 100. Thus, the local optimum values obtained by the local search methods are far from their global optimum.

5 Simulated Annealing

One problem with the use of local search is that the search algorithm may be trapped in a local optimum. The concept of the neighbourhood structure in the local search algorithm was then extended with the use of simulated annealing. Simulated annealing allows, in a limited way, transitions which correspond to an increase in the cost value.

This type of transition is controlled by an acceptance probability function and is depending on two parameters:
1. A measure of the progress of the search, which is indicated by a temperature variable which is set to a high value at the beginning of the search and gradually lowered as the search progresses
2. The magnitude of the increase in the cost function, which is simply the difference between the value of the current solution and the value of the candidate solution.

5.1 Acceptance Probability Function

The acceptance probability function is often formulated as an exponential function using Boltzmann distribution (Davis, 1987) (Osman & Potts, 1989) (Ogbu & Smith, 1990) (Van Laarhoven et.al., 1992) (Dowsland, 1993):

$$\exp\left(-\frac{\delta}{t}\right)$$

A linear function is sometimes used to approximate the distribution:

$$1 - \frac{\delta}{t}$$

where δ is the difference between the value of the current solution and the value of the candidate solution and t is the current temperature.

5.2   Generic Decisions

The algorithm uses an initial temperature of 100. Initial experiments indicated that this temperature is 'hot' enough to allow different areas of the search space to be explored and 'cold' enough to rapidly converge into local optimum.

The rate at which the temperature parameter is reduced is vital to the success of the annealing process. This is governed by two different parameters (Dowsland, 1993): the number of repetitions at each temperature and the rate at which the temperature is reduced. In this study, we use the cooling schedule proposed by Lundy and Mees (Lundy & Mees, 1986) which executes one iteration at each temperature and the temperature is reduced according to the following formula:

$$\alpha(t) = \frac{t}{(1 + \beta t)}$$

where β is a suitably small value

The value of $\beta = 0.0001$ is used in our study because initial experiments indicated that the value of 0.0001 gave a consistently good performance. For the purpose of comparison with the other heuristic methods, the algorithm terminates when 50,000 solutions have been generated.

The standard annealing process samples randomly from the neighbourhood of solutions. One potential problem with this sampling method is that some worse solutions may get selected before the better ones are sampled. In this study, we investigated another sampling method: cyclic sampling (Dowsland, 1993), to ensure that all neighbours are sampled once before any are considered for a second time.

5.3   Problem Specific Decisions

The solution space consists of n! different permutations of the n jobs to be scheduled. Two neighbourhood operators were evaluated and compared: exchange operator and move operator. The cost function used to evaluate a particular solution is the maximum tardiness value of all jobs in the schedule.

5.4   Preliminary Results

There were 3 experiments conducted to investigate the different acceptance probability functions, neighbourhood operators and sampling methods. During each run of the algorithm, the value of the best solution found was stored. Each experiment compared two or more techniques or parameters using 10 different runs on 10 different problems. The averages of these 10 runs were then used to plot the best solution against the number of solutions evaluated (see Appendix).

The algorithm uses an initial temperature of 100. and Lundy and Mees cooling schedule (Lundy & Mees, 1986). It terminates when 50,000 solutions have been generated.

Experiment 1 Boltzmann vs Approximation functions .
From this experiment we conclude that the approximation function is slightly better than the exponential function. Moreover, the approximation function is less computational expensive than the exponential function. Therefore, subsequent experiments used the linear approximation probability function.

Experiment 2 Exchange vs Move operators .
From this experiment, we conclude that the move operator performs slightly better than the exchange operator. Subsequent experiments used the move operator.

Experiment 3 Random vs Cyclic sampling methods .
The results of the experiment indicated that cyclic sampling does not improve the performance of the algorithm. In fact, it is worse than the random sampling method. Subsequent experiments used the random sampling method.

6   Tabu Search

Tabu search is another method which is capable of escaping from local optima by accepting inferior neighbouring solutions. This method considers all transitions both corresponding to an increase in the cost value and a decrease in the cost value, except for a set of prohibited or tabu transitions.

It uses a steepest descent method for moving from one solution to another. That is, a solution explores all of its neighbour solutions before replacing it with the lowest cost neighbour. If no better neighbour solutions exist, or the better ones are all tabu, then a worse solution is accepted.

This is a very slow process because $O(n^2)$ function evaluations are needed before each transition. To overcome this problem, a neighbourhood decomposition method is used (Reeves, 1993) (Glover & De Weera, 1993). The following describes how this is done (Reeves, 1993):

We examine the effect of exchanging or moving each jobs $\{J_1, ..., J_p\}$ where $p < n$, move to the best solution found and try moving jobs $\{J_{p+1}, ..., J_{2p}\}$ and so on. When we reach job J_n the search restarts from J_1.

This procedure is generalised by randomising the order in which the $p - job$ subsets were chosen, generating a random permutation of the numbers $\{1, ..., n\}$ whenever the $n'th$ job is reached, where n' is an integer multiple of n. This permutation then defines the order in which the n job moves will be explored.

6.1 Tabu List

The algorithm maintains a tabu list of solutions already traversed to avoid cycling. The tabu condition is stored as a complete solution. It means that a candidate schedule is tabu if it has the same job sequences as the one in the tabu list. This is a different approach because usually tabu conditions are identified in terms of their attributes (Taillard, 1990) (Laguna et.al., 1991) (Reeves, 1993) and not as complete solutions.

Since initial experiments indicated that larger tabu list size does not improve the performance of the algorithm, only 7 complete solutions are stored in the tabu list at any one time.

6.2 Preliminary Results

There were 2 experiments conducted to investigate the different neighbourhood sizes and neighbourhood operators. During each run of the algorithm, the value of the best solution found was stored. Each experiment compared two or more techniques or parameters using 10 different runs on 10 different problems. The averages of these 10 runs were then used to plot the best solution against the number of solutions (see Appendix).

The algorithm terminates when 200,000 solutions have been generated. Initial experiments suggested that exploring the whole neighbourhood before moving to a neighbourhood solution is a very slow process. Therefore, neighbourhood decomposition method is used to decompose the number of neighbours to be explored into a number of subsets.

Experiment 1 Neighbourhood size .
From this experiment we conclude that a subset size of 1 is superior to larger subset sizes. It means that the tabu search does not explore the neighbourhood before making a decision to move to a neighbour solution. Subsequent experiments used a subset size of 1.

Experiment 2 Exchange vs Move operators .
From this experiment we conclude that the exchange operator is superior to the move operator. Subsequent experiments used the exchange operator.

7 Genetic Algorithms

7.1 Chromosome Representation

In the work described here, a gene represents a job to be scheduled on the machine, has a fixed location in the chromosome and has an initially random number in a specified range as its value. An array of such genes makes up a chromosome. Each chromosome represents a valid schedule. A schedule is obtained by sorting the jobs represented by the genes in increasing order of their values.

This representation is not the most common representation used for scheduling problems. There are two main advantages of this representation over the more commonly used order based representation (McMahon & Hadinoto, 1993):

1. Genetic operators used for bit string representation such as one point crossover and uniform crossover can be used with this representation. Using these operators, there are no overheads incurred for preventing omission and duplication of genes

2. It ensures the separation between chromosomes and the solutions encoded ie. the genetic operators need not be aware of constraints imposed upon the chromosome representations. One such constraint is that a chromosome which represents a valid schedule must contain exactly one instance of each job

7.2 Population Module

The algorithm uses a random initialisation technique to generate the initial population of chromosomes. A value from the uniform distribution 1..n is generated for each gene of the chromosomes.

The algorithm uses a steady state deletion technique; ie. after each crossover the worst two chromosomes are deleted from the population. A crossover operation always generates two new chromosomes.

Three different parent selection techniques were evaluated: fitness is evaluation, rank based selection and using exponential function.

Fitness Is Evaluation. Using this technique, the probability of a chromosome being selected for crossover is proportionate to its fitness value. There are two potential problems with using the fitness value returned from the evaluation function (Davis, 1991):

1. The existence of a superior chromosome which is much better than its nearest competitor will cause rapid dominance of its genes in the population. The chromosome will have few chances to attempt recombination with the other population members and the population will probably experience rapid convergence.

2. The existence of several individuals with fitness values which are very close. This situation often arises at the end of the algorithm run. Without an increased amount of selection pressure, preserving good structures in chromosomes will just be a random walk through the population

Rank Based Selection (Whitley, 1989). Ranking acts as a function transformation which assigns a new fitness value to a chromosome based on its performance relative to other chromosomes. It intentionally discards information regarding the actual fitness value returned by the evaluation function and uses a relative value instead. A linear function is used to select parent chromosomes for reproduction such that the top ranked individual in the population is twice more likely to be selected than the median individual in the population.

Using Exponential Function. This technique uses a negative exponential function to inject bias towards good chromosomes in the population. A function is chosen which allocates a value of 100 to the best individual and a specified value for the median individual in the population. In this study we used a median value of 5.

7.3   Reproduction Module

The algorithm uses two genetic operators: uniform crossover and mutation. The uniform crossover uses randomly generated bit strings to determine from which parent a particular gene value is to be inherited. A value of 1 means that a gene value is inherited from one parent and a value of 0 means that it is inherited from the other parent. The mutation operator replaces a gene value with a new randomly generated value. This operator is controlled by a mutation rate parameter which determines the probability of a mutation on a particular gene.

7.4   Evaluation Module

To evaluate the fitness of an individual, its chromosome needs to be decoded into a feasible schedule. This is done by sorting the genes of the chromosome. The order of the jobs in the sorted chromosome defines the order of the jobs in the schedule. The tardiness value for each job is then calculated and the maximum of these values are taken to be the value of the schedule. The fitness value is obtained by subtracting the maximum tardiness from a constant (we used a constant value of 4,000).

7.5   Preliminary Results

An experiment was performed to investigate the different parent selection techniques used in the algorithm. During each run of the algorithm, the value of the best solution in the population was stored. There were 10 different runs on 10 different problems. The averages of these 10 runs were then used to plot the best solution against the number of generations (see Appendix).

Each gene had a random value between 0 and 999. These values, when sorted, were used to determine a valid schedule from the chromosome. It had an initial population of 500 chromosomes and a total of 250,000 chromosomes were generated. The algorithm used a mutation rate of 0.01.

Experiment Fitness vs Rank based vs Exponential .
Both exponential function and rank based selection converge rapidly in the first 50 generations. However, towards the end of the algorithm run, the exponential function is better than both fitness and rank based selection. Therefore, the exponential function with median of 5 is adopted as the parent selection technique for the genetic algorithm.

8 Performance Comparison

This experiment is performed to compare the four heuristic algorithms described in the previous sections, using their best techniques and parameters. During each run of the algorithm, the value of the best solution found was stored. Each experiment compared two or more techniques or parameters using 10 different runs on 10 different problems. The averages of these 10 runs were then used to plot the best solution against time (in seconds). The algorithms terminate when 60 seconds of CPU processing time elapsed.

From this experiment, we can conclude that for this class of scheduling problems simulated annealing is the best followed by tabu search, genetic algorithms and local search. The simulated annealing converges into the local optimum within 10 seconds of CPU processing time, the tabu search converges into the same value within 30 seconds of CPU processing time, while the genetic algorithm is very slow and does not converge within the 60 seconds of CPU processing time. As for the local search, its local optimum is very far from the global optimum, as compared to the other search algorithms.

9 Conclusions and Future Works

The most important conclusion derived from this research is that simulated annealing is the best heuristic search algorithm for single machine scheduling problems to minimise maximum tardiness. This is somewhat surprising, because simulated annealing is often considered as a very slow method. There are two possible explanations for this:
1. Simulated annealing works well with this kind of problem or solution space.
2. The other algorithms are not implemented as efficiently as they should be.

In recent years, genetic algorithms have gained popularity at the expense of other local search algorithms such as simulated annealing and tabu search. However, we have not found research that directly compare the performance of genetic algorithms with the other local search algorithms on this kind of problem. Most research into the use of genetic algorithms has concentrated on the use of different genetic operators and parent selection techniques. The work described here suggested that at least in the area of single machine scheduling problems, simulated annealing and tabu search are competitive. A similar result on different scheduling problems ie. job shop problems was reported by Vaessens et. al. (1994).

More investigations are needed to improve the performance of the genetic algorithms and tabu search and test them against more standard implementation of these methods. One possible extension to this work is to apply the algorithms to more complex problems such as multiple machines scheduling, multiple resources scheduling, etc.

Appendix

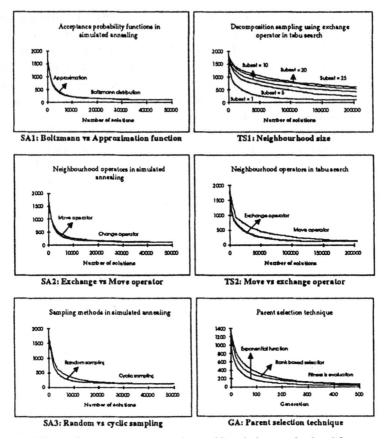

Fig. 1. Performance comparison of heuristic search algorithms

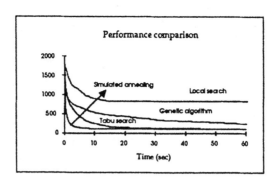

Fig. 2. Preliminary results.

References

1. Bagchi, S., Uckun, S., Miyabe, Y. and Kawamura, K.: Exploring problem specific recombination operators for job shop scheduling. Proceedings of the 4th International Conference on Genetic Algorithms (1991) 10–17
2. Baker, K.R.: Introduction to sequencing and scheduling. Wiley, New York (1974)
3. Brandimarte, P.: Neighbourhood serach-based optimization algorithms for production scheduling: a survey. Computer Integrated Manufacturing Systems 5 (1992) 167–176
4. Cleveland, G.A. and Smith, S.F.: Using genetic algorithms to schedule flow shop releases. Proceedings of the 3rd International Conference on Genetic Algorithms (1989) 160–169
5. Davis, L.: Job shop scheduling with genetic algorithms. Proceedings of the 1st International Conference on Genetic Algorithms (1985) 136–140
6. Davis, L.: Genetic algorithms and simulated annealing. Pitman Publishing, London (1987)
7. Davis, L.: Handbook of genetic algorithms. Van Nostrand Reinhold, New York (1991)
8. Davis, L. and Ritter, F.: Schedule optimization with probabilistic search. Proceedings of the 3rd Conference on Artificial Intelligence Applications (1987) 231–236
9. Dowsland, K.A.: Simulated annealing. Modern heuristic techniques for combinatorial problems, Blackwell Scientific Publications, Oxford (1993)
10. Fox, M. S.: Constraint directed search: a case study of job shop scheduling. Morgan Kaufmann Publishers, Los Altos, CA (1987)
11. French, S.: Sequencing and scheduling. Ellis Horwood, Chichester (1982)
12. Glover, F.: Tabu search - part 1. ORSA Journal on Computing 1 (1989) 190–206
13. Glover, F.: Tabu search - part 2. ORSA Journal on Computing 2 (1990) 4–32
14. Glover, F.: Tabu search: a tutorial. Interfaces 20 (1990) 74–94
15. Glover, F., Taillard, E. and de Werra, D.: A user's guide to tabu search. Annals of Operations Research 41 (1993) 3–28
16. Goldberg, D.E.: Genetic algorithms in search, optimisation and machine learning. Addison Wesley, Reading, Mass (1989)
17. Husbands, P., Mill, F and Warrington, S.: Genetic algorithms, production plan optimization and scheduling. Proceedings of the 1st International Conference on Parallel Problem Solving from Nature (PPSN) (1990) 80–84
18. Laguna, M., Barnes, J.W. and Glover, F.: Tabu search methods for a single machine scheduling problem. Journal of Intelligent Manufacturing 2 (1991) 63–73
19. Lundy, M. and Mees, A.: Convergence of an annealing algorithm. Math. Prog. 34 (1986) 111–124
20. McMahon, G. and Hadinoto, D.: A genetic algorithm for single machine scheduling problems. Working paper 1993-3-095/B, Bond University, Gold Coast, Australia (1993)
21. Ogbu, F.A. and Smith, D.K.: The application of the simulated annealing algorithm to the solution of the n/m/Cmax flowshop problem. Computers Operations Research 17 (1990) 243–253
22. Osman, I.H. and Potts, C.N.: Simulated annealing for permutation flow shop scheduling. Omega 17 (1989) 551–557
23. Reeves, C.R.: Improving the efficiency of tabu search for machine sequencing problems. Journal of Operational Research Society 44 (1993) 375–382

24. Rinnooy Kan, A.H.G.: Machine scheduling problem: classification. complexity and computation. Martinus Nijhoff, The Hague, Holland (1976)

25. Syswerda, G. and Palmucci, J.: The application of genetic algorithms to resource scheduling. Proceedings of the 4th International Conference on Genetic Algorithms (1991) 502–508

26. Taillard, E.: Some efficient heuristic methods for the flow shop sequencing problem. European Journal of Operational Research 47 (1990) 65–74

27. Vaessens, R.J.M., Aarts, E.H.L. and Lenstra, J.K.: Job shop scheduling by local search. Memorandum COSOR 94-05, Eindhoven University of Technology, The Netherlands (1994)

28. Van Laarhoven, P.J.M., Aarts, E.H.L. and Lenstra, J.K.: Job shop scheduling by simulated annealing. Operations Research 40 (1992) 113–125

29. Whitley, D.: The GENITOR algorithm and selection pressure: why rank based allocation of reproductive trials is best. Proceedings of the 3rd International Conference on Genetic Algorithms (1989) 116–121

30. Whitley, D., Starkweather, T., and Fuquay, D'A: Scheduling problems and travelling salesman: the genetic edge recombination operator. Proceedings of the 3rd International Conference on Genetic Algorithms (1989) 133–140

31. Widmer, M. and Hertz, A.: A new heuristic method for the flow shop sequencing problem. European Journal of Operational Research 41 (1989) 186–193

Encoding graphs for genetic algorithms: An investigation using the minimum spanning tree problem.

Pushkar Piggott[1] and Francis Suraweera[2]

[1] Department of Computer Science, University of Wollongong, Northfields Avenue, WOLLONGONG 2522, Australia.
[2] School of Computing and Information Technology, Griffith University, NATHAN 4111, Brisbane, Australia.

Abstract. We present a comparison of modified and unmodified GAs for the MST problem. A GA assembles successful gene substrings into improved solutions, but it does not have a mechanism to enforce global constraints. Graph theory contains many problems that are suitable for GAs because polynomial-time algorithms do not exist, but they often have global constraints. Special encodings and modifications of GA operators have been developed to deal with this difficulty.
We use the MST as an example problem because it is representative of the encoding difficulty, while the existence of polynomial-time algorithms makes the evaluation of performance relatively simple. We modify the GA crossover operator to preserve the property that an MST has 1 fewer edges than the number of vertices. Although this restricts the search space substantially, our results show that the expected benefits are not obtained. The GA demonstrates its power by successfully restricting the search without help.

1 Introduction

In this paper we present a comparison of modified and unmodified genetic algorithms (GAs) finding the the minimum spanning tree (MST) of a graph. Graphs are a versatile formalism that can be used to represent a wide variety of important and difficult real-world problems. Graph theory contains problems that are suitable for GAs because polynomial-time solutions for them do not exist. Encoding graph problems for GAs is known to be difficult however [1], and the encoding of the problem is critical to the success of a GA.

Finding the MST is a graph problem for which polynomial-time solutions do exist, and it is not our intention to compare the performance of GAs with these. We use the MST as a test problem because it allows us to check the quality of GA solutions for large numbers of randomly generated graphs within a reasonable time, and thus to compare the solutions obtained using different approaches.

GAs build global success from local success, but many graph problems are constrained by both local and global factors. In the MST example, a solution has exactly one fewer edges than the number of vertices in the graph. This constraint significantly limits the search space, but is not preserved by an unmodified GA.

We suggest a modification to the crossover operator that preserves the edge count in candidate solutions, and test it empirically. We find that the modification does not improve the performance of the algorithm. Instead, the initialization to the correct number of edges gives the unmodified version a substantial boost, and it is able to maintain its focus on candidates with the correct number of edges without the assistance of the modified operator.

We introduce the MST problem and GAs in Sections 2 and 3. We discuss the application of GAs to the MST problem in Section 4. Section 5 presents and discusses the experimental results, and Section 6 concludes the paper.

2   The Minimum Spanning Tree Problem

The MST problem has a venerable history in combinatorial optimization and computer science; Graham and Hell [2] give a good historical account of it. The problem was first formulated in 1926 by Otakar Boruvka who is said to have learned about it during the rural electrification of Southern Moravia. He gave a solution to find the most economical layout of a power-line network. The algorithms due to Kruskal [3], Prim [4], Dijkstra [5] and Sollin [6] are better known than Boruvka's early work, and they have been used as the basis for many other sequential and parallel algorithms.

Consider a finite undirected graph with a set of *vertices* V, a set of *edges* E connecting pairs of vertices from V, and a function W mapping each edge to a positive valued *weight*. The vertices and edges are sometimes referred to as *nodes* and *links*. The weight of a link may be thought of as its length, or the cost of traversing or implementing it. Such a graph can be formally defined as:

$$G = (V, E, W) \text{ with } m = |E| \text{ and } n = |V|$$

A *spanning tree* is a minimal set of edges from E that connects all the vertices in V into one *component*. If there is a path between every pair of nodes in G then G is said to be a *connected* graph, and at least one spanning tree can be found. A spanning tree contains no *cycles* since an edge can be removed from a cycle without disconnecting any vertices, and it has $n - 1$ edges. The weight of a spanning tree is the sum of the weights of its edges, and a *minimum spanning tree* is one of the spanning trees of G whose weight is minimal. In general a graph may have more than one MST, but a graph whose edges have distinct weights has just one.

Kruskal's algorithm maintains a list of edges sorted by weight, and selects edges starting from the lightest while avoiding cycles. Prim and Dijkstra's algorithms *grow* the tree from an arbitrary vertex by adding the least cost edge that extends the tree. The algorithms terminate when all the vertices are connected. Their time complexity derives from the sorting of the edges and the repeated checks for cycles and connected components [7].

3 Genetic Algorithms

Although some work in evolutionary programming preceded them, Holland and his colleagues at the University of Michigan pioneered the field of GAs in the mid 1960's [8]. Holland [8] and Goldberg [9] are excellent introductions to the field. The GA mechanism is a simplified simulation of natural selection in Darwinian evolution. It uses differential reproduction according to fitness, with genetic crossover, in a population of candidate solutions. A small amount of random mutation is also used. Holland established the mathematical basis of GAs [10] and showed that they are a near optimal approach to adaptation and search.

The GA approach can be described in terms of function optimization. Consider the case of a two input function. The input domains can be represented by axes in a cartesian plane, and the function can be thought of as specifying an altitude for each point on the plane. The result is a curved surface in three dimensions, and function optimization is the identification of the highest or lowest points. This model can be extended to functions of any finite number of inputs.

The traditional algorithm designer manipulates a formal description of the domains and operations of the function, exploiting their key properties to develop an efficient algorithm. The main tasks are the quest for key properties, and their effective combination and exploitation. This is exemplified by the traditional MST algorithms which exploit the key properties: relative edge weight, cycles and connected components. Functions used in artificial intelligence are often too complex for this algorithmic approach, and rule of thumb, or *heuristic*, knowledge about the behaviour of the function is used. The purpose is then to find an adequate solution that may not be a global optimum.

GAs require neither heuristics nor knowledge of key properties. The function is treated as a black box, and optimization is based on a sampling of points on its output curve. A simple hill climbing strategy relying on local shape is not adequate because many functions have numerous local maxima that would trap such a climber. GAs employ a population of candidate solutions chosen initially at random from the entire space of possible sets of inputs, and the problem of local minima is reduced because each currently superior candidate is combined with others to generate offspring candidates that are not necessarily located close to either parent.

For a GA to be effective, the quality of a candidate solution must be the result of the separate utilities of its component attributes. In reproduction the attributes are crossed with attributes from other good candidates by the *crossover* operation. If an offspring is a combination of superior attributes from both parents then these are preserved for future generations in the new individual whose own offspring test further recombinations. If an offspring is a combination of poor attributes then it does not survive, and no further effort is wasted. Further trials are allocated to more successful areas of the search space at the optimal exponential rate. GAs are perhaps unique in that this implicit parallelism allows them to use combinatorial explosion to their advantage. The number of schemata processed is generally quoted as proportional to the cube of the population, a more precise relationship is given in [11].

	De Jong	Schaffer *et al.*	Experiment
Population	50–100	30–59	100
Crossover Rate	0.60	0.75–0.95	0.66
Mutation Rate	0.001	0.005–0.01	0.001

Table 1. Values suggested in the literature, and the values used in the experiment.

A GA evaluates fitness, and applies fitness proportional reproduction with crossover and mutation to a population of candidate solutions. The GA designer is relieved of the burden of searching out and exploiting the key properties of the function to be optimized, but this burden is replaced by the need to choose from a wide variety of parameter values. The influence of various parameters in the context of variously shaped functions was studied by De Jong [9], and remains a central topic in GA research. De Jong tested a range of values for a selection of parameters on 5 functions whose different curves were hoped to characterize many functions which might arise in practice. His results have become part of the GA folk-lore. More recently Schaffer *et al.* [12] extended De Jong's work, concentrating on rapid convergence. They found some improvement could be gained from small populations with higher crossover and mutation rates. Table 1 compares values suggested by these two studies with the values used in our experimental work.

4 GAs and Graph Problems

The method used to encode the candidate solutions has a powerful influence on the effectiveness of a GA [1]. The structure of the encoding should match the structure of the candidate solutions such that the distinct attributes that are the building blocks of a solution are independent and can be mixed and matched. A binary string can be used to represent a membership function over a set of attributes, and this is generally accepted as the best approach [11]. Antonisse [13] contends that the apparent optimality of this method is an artifact of Holland's schema analysis however, and suggests that a more flexible analysis might show other methods to be as good.

The GA builds improved solutions by putting together the more successful gene substring "building blocks". Thus it builds global success from local success, without preserving global properties. For the MST problem, a bit string can represent a candidate solution by indicating which edges are used (Figure 1). The string represents a membership function over the edges in the graph, and the state of each bit determines whether the candidate includes the corresponding edge. Although we know that a correct solution will have exactly $n-1$ edges, this is a global property of the bit string that normal genetic crossover and mutation operators will not preserve. Such global properties of solutions are common in graph problems and contribute to their difficulty for GAs.

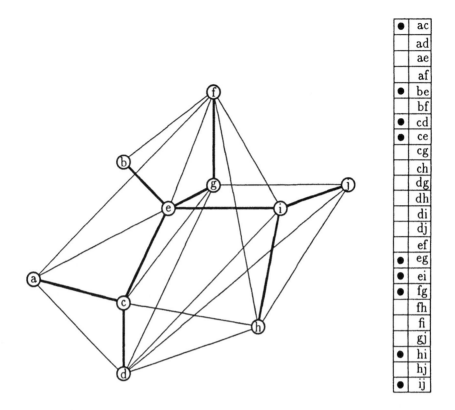

●	ac
	ad
	ae
	af
●	be
	bf
●	cd
●	ce
	cg
	ch
	dg
	dh
	di
	dj
	ef
●	eg
●	ei
●	fg
	fh
	fi
	gj
●	hi
	hj
●	ij

Fig. 1. A graph, its MST, and a bit string encoding for it.

To visualize the problem, recall that it can be modeled as a curve in a high dimensional space. With a bit string representation for trees, each point representing a candidate with $n-1$ edges is surrounded by ineligible candidates with too many or too few edges. The resulting curve resembles an echidna, or De Jong's test function F5 [9], which is accepted as one of the hardest to work with. There seems to be ample justification for abandoning the bit string representation and using an alternative that preserves the $n-1$ edge property. Such an encoding would restrict the algorithm to a much smaller search space, and we could hope for a considerable improvement in performance.

An examination of alternatives reveals that the bit string does have useful properties, however. The crossover operator requires an explicit ordering of the edges so that a single crossover point can be defined in both parents. Unfortunately, the edges of a graph have no such order. An alternative encoding, such as using $n-1$ vertex pairs to represent the edges in a candidate, would not define a relationship between the orderings of the edges in two candidates, and duplicate edges might occur in the offspring (Figure 2). By representing every edge explicitly, a bit string imposes an edge order, and allows crossover points to

				Crossover	Point				
ad^1	af^1	be^1	dh^1	di^1	ef^1	ei^1	fg^1	fi^1	Parent1

ac^2	af^2	bf^2	ce^2	cg^2	dh^2	fh^2	fi^2	hi^2	Parent2

ad^1	af^1	be^1	dh^1	di^1	dh^2	fh^2	fi^2	hi^2	Child

Fig. 2. An encoding using only $n-1$ gene locations does not prevent edge duplication.

be defined in the edge domain that are meaningful with respect to both parent strings.

Rather than use an encoding that enforces the desired global property, we can modify the crossover and mutation operators to preserve it. We describe such a modification in the next section and present experimental results that compare the performance of a GA with and without it. GAs using modified operators that take advantage of extra information about the function have been called *heuristic* GAs by Jog *et al.* [14]. By using such operators we encroach on the field of traditional algorithm design and lose the full generality of the generic GA mechanism.

If we define fitness solely in terms of edge weight then candidates with fewer edges will predominate, and the edgeless graph is optimal. *Connectivity* must also be a fitness criterion, and the MST problem can be described as the simultaneous minimization of both weight and number of components. The resulting algorithm is more general than the traditional algorithms because it does not distinguish between fully and partially connected graphs. It searches for the minimal weighted set of edges that spans as much of the graph as possible.

The relative importance of the two fitness factors is set by multiplying one by a bias constant before the two are added together to generate the overall fitness value. We subtracte one from the number of components, to give the number of edges missing, and multiply this by the weight limit set for edges. Halving the bias leads to more disconnected solutions, while performance degrades slightly if it is doubled.

5 An Experimental Comparison

Our modified crossover operator (Figure 3) does not establish the $n-1$ edge property, it only preserves it. We assume that the initial population consists of individuals with the correct number of edges, and then preserve this edge count between generations. An offspring inherits between 0 and $n-1$ edges from one parent, and the rest, up to $n-1$, from the other. A random crossover point is selected, the first parent's bit string is scanned, and edges are transferred to the offspring until the randomly chosen count is reached. The rest are transferred

parent1, *parent2* and *child* are bit strings of length m, containing $n - 1$ set bits.
i and c are randomly generated integers: $0 \leq i < m$ and $0 \leq c \leq n - 1$.

```
for edge := 1 to c do                 Copy c edges from parent1
    while parent1[i mod m] = 0 do      Skip edges absent from parent1
        inc(i);
    end;
    child[i mod m] := 1;
    inc(i);
end;

for edge := c + 1 to n - 1 do         Copy n - 1 - c edges from parent2
    while parent2[i mod m] = 0         Skip edges absent from parent2
    or child[i mod m] = 1 do           or already present in child
        inc(i);
    end;
    child[i mod m] := 1;
    inc(i);
end;
```

Fig. 3. A modified crossover operator that transfers $n - 1$ edges to the offspring.

from the second parent, omitting any already present. The modified mutation operator moves one edge, and thus avoids altering the number of edges.

Fitness is a continuous function yielding fractional values, but an individual must have a whole number of offspring. If individuals are chosen for reproduction stochastically, and the selection is weighted by fitness, then the integral number of reproductions reflect the fractional fitness values. We use rank proportional reproduction for our experimental system, but not stochastic selection. Instead, we assign fixed reproduction quotas to the higher ranks. This is a poor strategy and makes fine tuning of the relationship between rank and reproductive success impossible.

To qualify as an algorithm, a GA must include a mechanism for termination. We control the number of generations in a run by convergence. The algorithm terminates when a pre-determined fraction of the population has the same value. We found that this group of identical individuals did not necessarily include the best, suggesting that the mapping of rank to reproductive effectiveness could be improved. The reproduction quota system described above precludes finer tuning however, and we did not investigate further because our experiment is primarily concerned with other aspects of the algorithm.

We compared the modified and unmodified crossover operators on sets of randomly connected graphs. We used graphs with a edge to vertex ratio, or *connectivity*, of 5. Cheeseman *et al.* [15] show that with this connectivity there is a very high probability of a Hamiltonian Circuit in a graph, but that the computational cost of finding it is also high. At lower connectivities the graph is likely to be broken into a number of discrete components, and at higher connectivi-

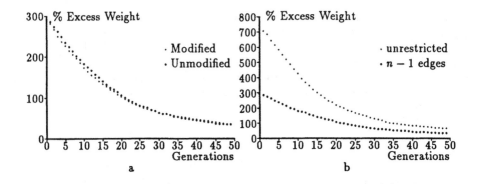

Fig. 4. Comparisons of: a) GAs using modified and unmodified crossover operators, and b) the unmodified GA with unrestricted and $n-1$ edge count initial populations.

ties circuits become increasingly easy to find. This result suggests that graphs with a connectivity of 5 are also the most challenging for a search for spanning trees, although finding *minimum* spanning trees may remain difficult at higher connectivities.

The modified operator does not establish the $n-1$ edge property, and the modified algorithm therefore requires an initial population of candidates with the correct number of edges. The same population of candidate solutions was used to initialize both versions of the algorithm, to allow a valid comparison shown in Figure 4a. Figure 4b reveals the effect of this constrained initial population by comparing the same results for the unmodified algorithm with the outcome using unrestricted initial populations.

Each data point shows the excess weight of the best candidate at a given generation as a percentage of the true MST weight calculated by conventional means. Each curve is averaged over 100 trials using different graphs.

The graphs reveal that the solutions found by the GAs are not perfect, the average final solution is nearly 40% over-weight. The performance could perhaps be improved, but this is not our focus. The GA cannot compete with the performance of purpose-built polynomial-time algorithms, and our emphasis is on the effect of the modification to the crossover operator.

The modification limits the search to the subspace consisting of candidates having the correct number of edges. This provides a very significant reduction in search space size. Figure 4a shows that this does not yield an improvement in performance, however. The unmodified algorithm loses a little ground in the early stages, but catches up later. Although the unmodified crossover introduces offspring with the wrong number of edges, the graph demonstrates that the search is successfully controlled by the disconnectivity penalty alone. The superior performance of the unmodified algorithm in the later stages suggests that it exploits the greater diversity and freedom of movement through the search space that the lack of restriction allows it to discover better solutions.

Contrast the small variation in performance in Figure 4a with the substantial effect, shown in Figure 4b, of providing a controlled initial population. The average unrestricted candidate is heavier than the average candidate with $n-1$ edges because n is much smaller than m. The GA does a good job of removing heavier candidates, but then converges before it can reach solutions as good as those generated from a restricted initial population.

Modifying the algorithm clearly does not have the intended beneficial effect, and in this case greater benefit is obtained merely by selecting the initial population. This example demonstrates both that the unmodified GA is very robust, and that experience from traditional algorithm design may be misleading when applied to a GA. Although surprising, it is gratifying to find that the unmodified GA has such strengths.

6 Conclusion

In this paper we put forward the suggestion that global constraints on solutions, which are common in graph problems, motivate altering the encoding used for candidate solutions. However, we found that the bit string encoding has valuable properties, and took the alternative course of modifying the crossover and mutation operators to preserve the desired constraint.

We found that the modification failed to deliver the substantial improvement it promised, and that a simple constraint on the initial population does yield an improvement. Our experimental study vindicates the use of simple empirical tests to check the effects of modifications to the GA. The modifications may be unnecessary, and simple constraints on the initial population may be more effective.

References

1. De Jong, K.A., Spears, W.A.: Using genetic algorithms to solve NP-complete problems. Proc. 3rd Annual Conf. on GAs. Morgan Kaufmann San Mateo (1989)
2. Graham, R.L., Hell, P.: On the history of the minimum spanning tree problem. Annals of the History of Computing. 7 (1985) 43–57
3. Kruskal Jr., J.B., On the shortest spanning subtree of a graph and the traveling salesman problem. Proc. ACM 7(1) (1956) 48–50
4. Prim, R.C.: Shortest connection networks and some generalizations. Bell Systems Tech. J. 36 (1957) 1389–1401
5. Dijkstra, E.W.: A note on two problems in connection with graphs. Numerische Mathematik 1 (1959) 269–271
6. Sollin, M.: Le trace de canalisation. Berge, C., Ghouilla-Houri, A. eds.: Programming, Games, and Transportation Networks. John Wiley NY (1965)
7. Baase, S.: Computer Algorithms: Introduction to Design and Analysis. Addison-Wesley Reading MA (1988)
8. Holland, J.H.: Genetic algorithms. Scientific American July (1992) 44–50
9. Goldberg, D.E.: Genetic Algorithms in Search, Optimization, and Machine Learning. Addison-Wesley Reading MA (1989)
10. Holland, J.H.: The dynamics of searches directed by Genetic Algorithms. In Y. C. Lee, ed.: Evolution, Learning and Cognition. World Scientific, Singapore (1988) 111–127

11. Goldberg, D.E.: Sizing populations for serial and parallel genetic algorithms. Proc. 3rd Annual Conf. on GAs. Morgan Kaufmann San Mateo (1989) 124–132
12. Schaffer, J.D., Caruana, R.A., Eshelman, L.J., Das, R.: A study of control parameters affecting online performance of genetic algorithms. Proc. 3rd Annual Conf. GAs. Morgan Kaufmann San Mateo (1989) 51–60
13. Antonisse, J.: A new interpretation of schema notation that overturns the binary encoding constraint. Proc. 3rd Annual Conf. on GAs. Morgan Kaufmann San Mateo (1989) 86–91
14. Jog, P., Suh, J.H., Gucht, D.V.: The effects of population size, heuristic crossover and local improvement on a genetic algorithm for the traveling salesman problem. Proc. 3rd Annual Conf. GAs. Morgan Kaufmann San Mateo (1989) 86–91
15. Cheeseman, P., Kanefsky, B., Taylor, W.M.: Where the really hard problems are. Proc. 12th Int. J. Conf. on AI. Morgan Kaufmann San Mateo (1991) 331–337

Lecture Notes in Artificial Intelligence (LNAI)

Lecture Notes in Computer Science